A FORMAL APPROACH TO HARDWARE DESIGN

THE KLUWER INTERNATIONAL SERIES IN ENGINEERING AND COMPUTER SCIENCE

VLSI, COMPUTER ARCHITECTURE AND DIGITAL SIGNAL PROCESSING

Consulting Editor
Jonathan Allen

Other books in the series:

HOT-CARRIER RELIABILITY OF MOS VLSI CIRCUITS, Y. Leblebici, S. Kang
ISBN: 0-7923-9352-X
MOTION ANALYSIS AND IMAGE SEQUENCE PROCESSING, M. I. Sezan, R. Lagendijk
ISBN: 0-7923-9329-5
HIGH-LEVEL SYNTHESIS FOR REAL-TIME DIGITAL SIGNAL PROCESSING: The Cathedral-II Silicon Compiler, J. Vanhoof, K. van Rompaey, I. Bolsens, G. Gossens, H. DeMan
ISBN: 0-7923-9313-9
SIGMA DELTA MODULATORS: Nonlinear Decoding Algorithms and Stability Analysis, S. Hein, A. Zakhor
ISBN: 0-7923-9309-0
LOGIC SYNTHESIS AND OPTIMIZATION, T. Sasao
ISBN: 0-7923-9308-2
ACOUSTICAL AND ENVIRONMENTAL ROBUSTNESS IN AUTOMATIC SPEECH RECOGNITION, A. Acero
ISBN: 0-7923-9284-1
DESIGN AUTOMATION FOR TIMING-DRIVEN LAYOUT SYNTHESIS, S. S. Sapatnekar, S. Kang
ISBN: 0-7923-9281-7
DIGITAL BiCMOS INTEGRATED CIRCUIT DESIGN, S. H. K. Embadi, A. Bellaouar, M. I. Elmasry
ISBN: 0-7923-9276-0
WAVELET THEORY AND ITS APPLICATIONS, R. K. Young
ISBN: 0-7923-9271-X
VHDL FOR SIMULATION, SYNTHESIS AND FORMAL PROOFS OF HARDWARE, J. Mermet
ISBN: 0-7923-9253-1
ELECTRONIC CAD FRAMEWORKS, T. J. Barnes, D. Harrison, A. R. Newton, R. L. Spickelmier
ISBN: 0-7923-9252-3
ANATOMY OF A SILICON COMPILER, R. W. Brodersen
ISBN: 0-7923-9249-3
FIELD-PROGRAMMABLE GATE ARRAYS, S. D. Brown, R. J. Francis, J. Rose, S. G. Vranesic
ISBN: 0-7923-9248-5
THE SECD MICROPROCESSOR, A VERIFICATION CASE STUDY, B. T. Graham
ISBN: 0-7923-9245-0
HIGH LEVEL SYNTHESIS OF ASICs UNDER TIMING AND SYNCHRONIZATION CONSTRAINTS, D. C. Ku, G. De Micheli
ISBN: 0-7923-9244-2
FAULT COVERING PROBLEMS IN RECONFIGURABLE VLSI SYSTEMS, R. Libeskind-Hadas, N. Hassan, J. Cong, P. McKinley, C. L. Liu
ISBN: 0-7923-9231-0
VHDL DESIGNER'S REFERENCE, J-M. Bergé, A. Fonkoua, S. Maginot, J. Rouillard
ISBN: 0-7923-1756-4
A FORMAL APPROACH TO HARDWARE DESIGN, J. Staunstrup
ISBN: 0-7923-9427-5

A FORMAL APPROACH TO HARDWARE DESIGN

by

Jørgen Staunstrup
Department of Computer Science,
Technical University of Denmark

KLUWER ACADEMIC PUBLISHERS
BOSTON / DORDRECHT / LONDON

A C.I.P. Catalogue record for this book is available from the Library of Congress.

ISBN 0-7923-9427-5

Published by Kluwer Academic Publishers,
P.O. Box 17, 3300 AA Dordrecht, The Netherlands.

Kluwer Academic Publishers incorporates
the publishing programmes of
D. Reidel, Martinus Nijhoff, Dr W. Junk and MTP Press.

Sold and distributed in the U.S.A. and Canada
by Kluwer Academic Publishers,
101 Philip Drive, Norwell, MA 02061, U.S.A.

In all other countries, sold and distributed
by Kluwer Academic Publishers Group,
P.O. Box 322, 3300 AH Dordrecht, The Netherlands.

Printed on acid-free paper

Printed in the Netherlands

DEDICATION

To Per Brinch Hansen

for setting course

CONTENTS

PREFACE

This is a book about designing computations to be realized by application specific hardware. Today, such hardware is often complicated and its cost is dominated by the design effort. This has for a long time been the case for software development. Therefore, there is an increasing similarity between hardware and software design, and they share a need for powerful, reliable, and efficient design methods. In software design there has been a trend towards using of high-level languages, encouraging the use of abstraction and rigorous analysis methods. This book supports a similar trend for hardware design by introducing a high-level design language, called Synchronized Transitions. The aim is to reduce the design time/effort by shifting as many tasks as possible from low-level circuit oriented models to high-level models. Design descriptions in Synchronized Transitions provide models of computations which can be utilized for many different purposes, for example:

- experimentation,
- verification, and
- performance analysis,

Furthermore, the descriptions can be refined to obtain circuit realizations, either through automatic synthesis or by manually guided transformations.

It is difficult to overestimate the benefits of making an abstract model of a design before proceeding with low-level implementation details. Nevertheless, omitting this is a very common mistake that many of us have made (more than once). By making an abstract model, the designer is able to examine a design on a high level of abstraction early in the design process. The models created using Synchronized Transitions enable the designer to perform different kinds of analysis and verification based on descriptions in a *single* language. It is, for example, possible to use *exactly the same design description* both for formal verification and synthesis.

This book shows the usefulness of a formal approach without being too dogmatic about its use. The rigor provided by formalism is needed for handling the complicated parts of a design. However, there are also parts where formal methods have no role to play. One should not insist that everything is done formally, but leave it to the designer's good taste to decide when formal analysis is appropriate and when more informal approaches are adequate. The intention with Synchronized Transitions is to provide a language where formal analysis is a possibility, but not mandatory. Training is a prerequisite for using any approach or tool, including formal methods. By reading this book and solving some of the problems, a designer should get a sufficient background for using abstract models and a formal approach to solve complicated design problems.

A circuit has a high potential for parallelism; all parts may operate simultaneously and having it do so is no more expensive than having just one part active at a time. To fully exploit this potential, the models created with Synchronized Transitions encourage parallelism. Traditionally, circuit design is based upon a synchronous model of parallelism, where all parts operate in lock-step. This is a well understood and well developed model at the circuit level. However, when used at a high-level it often means that the timing details of the low-level realization are reflected in the high-level description. Most of the designs discussed in this book attempt to avoid this by using an asynchronous model of parallel computation where behavior is separated from timing.

Synchronized Transitions is intended for designing application specific integrated circuits (often called ASICs). Such circuits are usually designed for a particular application, in contrast to general purpose circuits, as for example a microprocessor. For such specialized circuits the design process is often the bottleneck in the development. On the other hand, ASICs may not require the same optimal performance needed in mass-produced products such as memory chips or microprocessors. Furthermore, an ASIC is often just one part of a product with many other hardware and software components. For such products, the design time is usually critical which again is a reason to stress the importance of using high-level models.

Using the Book for Teaching

This book gives an in depth introduction to a particular formal design approach. There is a collection of exercises at the end of each chapter. Most of these are small exercises aimed at training the material presented in that chapter.

The book can be read by students or practioners without prior knowledge about formal designs methods. To follow the derivations in detail and to solve the exercises require a good command of elementary logic and either a programming language or a hardware description language.

The material has been used for a one semester course (14 weeks) on formal methods given at the Technical University of Denmark. The course consisted of lectures covering all chapters supplemented by a weekly training session where students discussed solutions to a selection of the exercises presented at the end of each chapter. The training sessions were also used for introducing the tools supporting Synchronized Transitions.

It is also possible to use the book for self study, either by reading it to understand the possibilities and limitations of using a formal approach, or by solving some of the exercises to get an in-depth understanding.

The last section of each chapter gives the background for the material introduced in the chapter. These sections contain references to some of the original publications, together with an overview of previous or alternative research work. The final chapter, the epilog, gives a personal summary of the most important achievements and shortcomings of the formal approach described in the remaining chapters of the book.

Tools

Synchronized Transitions is supported by a collection of prototype tools some of which are described in this book, in particular the translator ST2LP. All the tools are written in C and runs under Unix. They are available free of charge, and without a license, by anonymous ftp from the publisher: world.std.com, ftp.wkap.nl, from the Technical University of Denmark: ftp.id.dth.dk, or by contacting the author: jst@id.dth.dk. In addition to the tools, one can get a complete up-to-date reference manual and other documentation, including solutions to the exercises.

LP is written in CLU and runs under Unix. Release 2.4 is currently available free of charge, and without a license, by anonymous ftp from larch.lcs.mit.edu. Prospective users can retrieve an executable version of LP, along with a supporting run-time library containing sample axiom sets and proofs, for DECstations, Sparcstations, MIPS machines, Vaxes, or Sun-3 workstations, and soon for DEC Alpha. Source code is also available.

Acknowledgements

Anders P. Ravn participated in shaping the fundamental concepts of Synchronized Transitions, and a number of people have assisted in its further development. Christian D. Nielsen, Niels Maretti, Niels Mellergaard, Henrik Hulgaard, Per Henrik Christensen, and Jacob Christensen have all contributed with improvements of the notation, design methods, and tools. The work on mechanical verification has been done in cooperation with John Guttag, Steve Garland, Niels Maretti, and Niels Mellergaard. Niels Mellergaard also wrote the translator that forms the basis of all currently available tools supporting Synchronized Transitions. Michael Kishinevsky assisted in formulating the Muller model of speed-independent circuits that has strongly influenced the chapter on self-timing. The hand drawing shown in the introduction of chapter 7 is made by Jan Madsen who also gave valuable comments on various parts of the manuscript.

Niels Maretti and Niels Mellergaard have read earlier versions of this manuscript and suggested numerous improvements. Bo Stig Hansen has resolved a number of problems related to the typesetting and printing.

Last but not least, Mark Greenstreet has contributed to all aspects of the development of Synchronized Transitions. His enthusiasm and creativity have been a constant source of inspiration. I am very grateful to Mark for many stimulating discussions over a period of almost ten years during which Synchronized Transitions has evolved.

Financial support for developing Synchronized Transitions has been given by Århus University, The Danish Technical Research Council, Digital External Research, Mogens Balslev's fond, and The Danish Natural Science Research Council.

1

FORMAL DESIGN METHODS

Formal methods can be very powerful tools for solving complex problems. In chemistry, for example, it is possible to describe, predict, and understand many complicated phenomena by using molecular formulas, e.g., H_2O, and transformations such as $2Na + 2H_2O \rightarrow 2NaOH + H_2$. A formal method is such a combination of a rigorously defined symbolic notation and a collection of well-defined transformations. Formal methods are used in many fields; often these are rooted in mathematics, but this is not always the case, as the example from chemistry shows. Some people may find this use of the word formal too liberal. Maybe they associate the word with something very remote and theoretical. This is not how the term is used here. As a contrast, consider the following example of a description that is not formal. It is taken from a cookbook with fish recipes [56]:

> *Baked Bream (Daurade rôti)* – Lay strips of fat bacon and anchovy fillets over the fish. Pack it into oiled parchment and bake it in a baking dish. When it is done, add a little white wine to the baking dish, boil it down and make it into a smooth sauce with butter. [1]

Clearly this recipe is usable only if you are already an experienced cook. It does not say anything about how long the fish should be cooked, nor does it say anything about the quantities of any of the ingredients. This particular cookbook has 967 other similar recipes. However, most people would need much more precise instructions to prepare this dish.

[1] The translation from Danish to English was done by Robin Sharp.

This book describes an approach to circuit design based on formal descriptions and transformations. However, the main point is not the use of formal notation or rigorous manipulations, but providing appropriate abstract models that enable a designer to describe and analyze his ideas without having to construct a low-level realization. It should be stressed that, this approach (and other recently proposed formal methods) is based on limited use, and it still lacks some of the elegance, maturity, and general acceptance of the formal methods employed in other fields.

1.1 WHY USE FORMAL METHODS?

Formal methods have gained widespread acceptance in many fields, for example, chemistry; but why are they needed in integrated circuit design? For exactly the same reasons they are used in other fields; they enable us to work on a high level of abstraction. With today's technology, it is possible to design circuits with millions of components, but to master this, it must be possible to abstract from the details. The primary commercial force behind formal methods is a wish to reduce the design time and hence shorten the time to develop new products. In addition to this, there are several other important motivations:

- Design of integrated circuits is quickly changing *from being an art practiced by a few to a routine task performed by many*. This necessitates good design techniques, in particular notations and tools for capturing and documenting a design that can be handled by designers who do not have intimate knowledge about semiconductors, fabrication processes, propagation delays, etc.
- In the natural sciences, formalization is a way to *capture and document models*. Experimentation and informal reporting are also important steps, but the goal is usually to find simple and concise formulas relating key quantities. Again, chemistry is a good example. A milestone in this field was the realization that all molecules can be described as combinations of around 100 chemical elements. Molecular formulas are based on this insight, and without it, the formulas would not be so useful. The abstract formulas themselves are not enough. The right insight or model is needed as the basis for a simple and powerful formal method. The approach described in this book models a circuit as a concurrent computation.

- Today many integrated circuits are designed as part of a project creating both hardware and software components. It is, therefore, desirable that the same or very similar methods are used to design both hardware and software. Formal methods have had significant impact on software development, a similar trend is emerging for hardware design.
- Integrated circuits are used in a wide variety of products, some of which are safety critical like cars, airplanes, and traffic control; others are consumer products expected to work perfectly for many years (without updates, repairs, or other maintenance). Such applications require design methods that yield *reliable products at a reasonable cost*. Formal methods have a role to play in the design of such products, because they enable us to verify a design at a very early stage of the development. It is, for example, cheap to assure against many design errors using formal verification techniques.

However, formal methods alone are not enough. There are many examples of problems they cannot guard against, for example, fabrication faults or an unfriendly user interface. Therefore, formal methods are only a supplement to a number of other techniques used by the designer.

1.2 MODELS OF INTEGRATED CIRCUITS

Formal notation and rigorous transformations are not by themselves a sufficient basis for a successful formal method. The abstract models on which the formal method is based are of equal importance. Before introducing any formal notation a few alternative models of integrated circuits are considered. This leads to the model used in This book, where the behavior of an integrated circuit is described as a concurrent computation.

Consider a circuit controlling a simple traffic light placed at the intersection of two roads called NS (North South) and EW (East West) (see figure 1.1). Sensors make it possible to detect whether cars are waiting, and the light is supposed to change in the direction of waiting cars. There are many ways to design such a traffic light controller; however, it is an indispensable requirement that the light is safe, i.e., that it is *always* red in one of the two directions. So it must be required that

$$NSlight = red \;\; OR \;\; EWlight = red$$

The design approach introduced in this book provides the designer with rigorous techniques (and supporting tools) to ensure such safety properties.

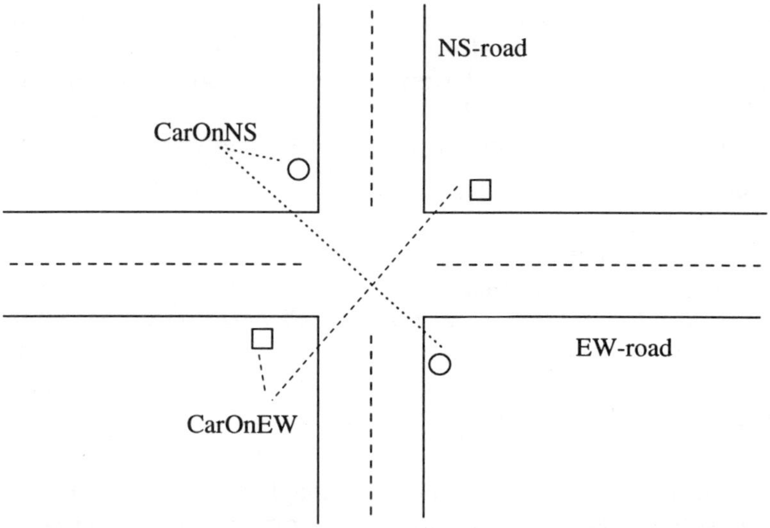

Figure 1.1 Highway intersection with a traffic light

A formal method is based on an abstract symbolic notation where the meaning of formulas is rigorously defined. Therefore, many formal methods employ mathematical equations. However, as the example with the molecular formulas shows, mathematical equations are not the only possibility. In most introductory texts on VLSI design, one finds mathematical equations such as this (taken from [50]):

$$I_{ds} = \frac{\mu\epsilon W}{2LD}(V_{gs} - V_{th})^2$$

Such equations are valuable for describing and understanding the operation of the semiconductor devices making up a VLSI chip. They are, however, inadequate for designing a complete chip, because they focus on the microscopic electrical details instead of the macroscopic functional behavior. It is, for example, impractical to use equations such as the one just shown to verify that the traffic light is safe.

A good formal method helps the designer grasp and maintain an overview of the intended computation. Within software design, numerous high-level languages have been developed to describe computations, and recently many attempts have been made to utilize such high-level programming languages to describe integrated circuits. One such attempt is the MODEL language used in the SOLO design systems (from European Silicon Structures). The language

```
PART Adder (n) [a(0:n), b(0:n), cin] -> sum(0:n), cout
   SIGNAL c(0:n+1)
   INTEGER i
   cin -> c(0)
   FOR i = 0:n CYCLE
      Add1[ a(i), b(i), c(i) ] -> sum(i), c(i+1)
   REPEAT
   c(n+1) -> cout
END
```

Figure 1.2 An adder described in MODEL

has some similarities with Pascal. Figure 1.2 shows an adder. The first line describes the interface (external signals) to the adder. Next, the internal (carry) signals are declared. The body describes the structure of the adder, i.e., which parts it consists of and how these are connected. For example, cin is connected to $c(0)$.

Although this description is written in a high-level language, it does not capture the dynamic computation (behavior) of the adder. It is an abstract description of how to assemble the chip, i.e., a structural description. Such structural descriptions contain details that are irrelevant in the early phases of a design where one is mainly concerned with the high-level properties of a design. In contrast to this, a behavioral description specifies the computation without indicating how the circuit is constructed. In the early phases of a design, such behavioral descriptions must be preferred, because they allow the designer to concentrate on *what* the circuit does, without worrying about *how* it is done.

Figure 1.3 shows a part of a behavioral description (taken from the documentation of a large VLSI project). It describes the computation of 128 sums and 64 comparisons. However, the program text prescribes that this must be done using 32 iterations, each of which consists of computing four sums ($sumA - sumD$). These are then compared to yield new values of TM. So this program specifies that the computation consists of 192 steps executed one after the other. However, the real intention behind the design is to specify a computation consisting of 64 elements computing in parallel. Such *parallel computing structures* are essential in an efficient integrated circuit. It is, therefore, important that parallelism can be expressed explicitly and enables the designer to control which computations are performed concurrently. This

```
...
FOR i = 0 TO 31 DO BEGIN
   sumA = TM[i] + F(i);
   sumB = TM[i+32] + FI(i);
   sumC = TM[i] + FI(i);
   sumD = TM[i+32] + F(i);
   TMnew[2i] = min(sumA, sumB);
   TMnew[2i+1] = min(sumC, sumD);
END;
...
```

Figure 1.3 Sequential description of summation

book introduces a design style based on such a parallel model of computation, where it is possible to express the parallelism needed in the example shown in figure 1.3. A parallel model encourages the designer to explicitly describe the parallelism of his design. It turns out that this parallelism also facilitates formal verification, because the parts executed in parallel can be verified independently of each other. The next section sketches a language (formal notation) for describing such parallel computations.

1.3 SYNCHRONIZED TRANSITIONS

This section describes a few key elements of a language for circuit design, called Synchronized Transitions. The notation is the basis for the formal approach described in this book. A design is modeled as a (usually large) set of independent transitions. For example:

$$\ll carry := a0\ AND\ b0 \gg$$

This transition describes a computation where a state variable *carry* is updated and becomes the logical AND of *a0* and *b0*. This update is done repeatedly, so every time *a0* or *b0* change, the value of *carry* may change. To describe a nontrivial circuit, a large number of transitions are needed. Each transition corresponds to an independent subcircuit executed in parallel with all the other transitions.

Consider a simple ripple-carry adder that adds two 2-bit registers, *a*, *b*, and delivers the result in a register, s. The bits of these registers are denoted $a0, a1, b0, b1, s0, s1$, and $s2$, where the 0'th bit is the least significant. Furthermore, register c contains the carry bits.

$$\ll c1 := a0 \ AND \ b0 \gg$$
$$\ll s0 := a0 \neq b0 \gg$$
$$\ll s1 := (a1 \neq b1) \neq c1 \gg$$
$$\ll s2 := (a1 \ AND \ b1) \ OR \ (a1 \ AND \ c1) \ OR \ (c1 \ AND \ b1) \gg$$

This describes four transitions; the last three show how the bits of the sum are computed. All four transitions are concurrent, and they execute repeatedly (just like the four subcircuits they describe). This means that the sequence in which new values appear in $s1$, $s2$, and $c1$ cannot be predicted. Hence, $s2$ may change value more than once before it stabilizes at the correct value. This stabilization will progress from the least significant bits $(s0, c1)$ to the most significant. Exactly the same thing happens in the circuit. For a programmer, it may seem odd that the circuit computes these spurious values before stabilizing. This can be avoided by describing the computation sequentially, but then it would no longer be an adequate abstraction of the circuit. Therefore, *the sequencing is deliberately not specified*, to reflect the operation of the circuit. A similar approach has been suggested in [10] for algorithm design.

Example: Scheduling a Meeting

This problem and its solution are presented in the book *Parallel Program Design* by Chandy and Misra [10]. Although the example is not oriented towards circuit design, it is a nice illustration of the computational model and design style advocated in this book. The problem is to find the earliest meeting time acceptable to every member of a group of people. Time is represented by an integer and is nonnegative. To keep the notation simple, assume that the group consists of three people: F, G, and H. Associated with persons F, G, H are functions f, g, h (respectively), which map times to times; $f(t)$ is the earliest time at or after t when person F can meet. Hence, for any t, person F can meet at time $f(t)$ and cannot meet at any time u where $t \leq u < f(t)$. The meanings of g and h are analogous. For example:

t	8	9	10	11	12	13	14	15	16	17	18
$f(t)$	9	9	10	13	13	13	16	16	16	17	20

In this example, F can meet at 9 or 10, but not at 11. The earliest F can meet after 11 is 13. It is assumed that there is at least one possible meeting time for

F, G, and H. Let t denote an estimate of the meeting time, if $f(t) > t$ then F cannot meet at t and $f(t)$ indicates the earliest time after t when F can meet; hence the estimate of the meeting time should be updated. This is expressed more succinctly by the transition

$$\ll f(t) > t \text{ -> } t := f(t) \gg$$

Similar transitions for G and H express their local updating of the estimate t. By combining the following three transitions, one obtains a computation that produces the earliest time at which F, G, and H can meet (after or at the time given by the initial value of the state variable t):

$$\ll f(t) > t \text{ -> } t := f(t) \gg$$
$$\ll g(t) > t \text{ -> } t := g(t) \gg$$
$$\ll h(t) > t \text{ -> } t := h(t) \gg$$

The state variable t is repeatedly increased by one of the three transitions until t gets a value such that

$$t = f(t) = g(t) = h(t)$$

which by definition of f, g, and h is a possible meeting time. Note the chaotic nature of this computation: each person updates the meeting time independently of the others. However, no potential meeting time is missed by one person making an excessive update. It might happen that one person updates the meeting time in small increments, whereas one of the others could make a larger increment. As was the case with the adder, there is no attempt to avoid such superfluous operations by sequencing the computation in a clever way. In many circuits, it would be less efficient to try to avoid these superfluous operations; the adder is a good example. This is not always the case, but then the optimization is a separate concern, left for a later refinement of the design.

Example: A Traffic Light Controller

Consider again the traffic light controller introduced in section 1.2. It is quite simple to express the design of such a controller in Synchronized Transitions. For example, when the light is green in the north-south direction and a car going east-west approaches the intersection, the light will change:

$$\ll (NSlight = green) \ AND \ CarOnEW \text{ -> } NSlight := yellow \gg$$

If the light is green in the east-west direction, then cars may enter the intersection from the east-west road, and likewise, for the north-south road:

$$\ll EWlight = green\ AND\ CarOnNS\ {-}{>}\ EWlight := yellow \gg$$
$$\ll EWlight = yellow\ {-}{>}\ EWlight, NSlight := red, green \gg$$
$$\ll NSlight = green\ AND\ CarOnEW\ {-}{>}\ NSlight := yellow \gg$$
$$\ll NSlight = yellow\ {-}{>}\ NSlight, EWlight := red, green \gg$$

There are four state variables, $EWlight, NSlight, CarOnNS$, and $CarOnEW$. The first two can take the values $red, yellow$, or $green$, and the last two are booleans that can be $TRUE$ or $FALSE$. The four transitions describe how the state variables change values; for example, when the condition

$$EWlight = green\ AND\ CarOnNS$$

holds, then $EWlight$ may change to $yellow$. Hence, the above describes a (finite) state machine. In chapter 2 the notation is described in further detail.

1.3.1 Verification

A formal description of a design like the one shown above for the traffic light is an important step in the design, because all details must be explicitly specified. However, a description by itself is only of limited value unless it can be used as the basis for other steps of the design process like verification and refinement. For those properties of the circuit that can be described rigorously, it is possible to do a formal verification (a mathematical proof) that the formal description has those properties. When this is possible, it is the fastest and cheapest way of verifying the design, because errors are revealed very early in the design process. The traffic light controller is an illustration of this.

To prevent collisions, the light must always be red in one of the two directions. This requirement is expressed with the assertion

$$EWlight = red\ OR\ NSlight = red$$

Such an assertion is called an invariant and it can be verified that all computations corresponding to the design description satisfy the invariant. In this simple example, it is feasible to enumerate and check all possible states; however, in general this becomes very laborious and in chapter 3 an alternative verification technique is introduced. Using this, it is shown that all transitions maintain the invariant by checking them one at a time. Using this technique, the verification effort is independent of the size of the state space or length of the computation.

This is an example of the possibilities offered by formal design methods, where it is possible to check that a design has certain desirable properties, expressed as invariants, at a very early stage of the design process. The traffic light example also illustrates an obvious limitation of formal methods: it is not possible to give an absolute guarantee against cars colliding. The drivers may, for example, not obey the traffic laws or the light may malfunction because of a power shortage. Considerations at the formal level are always based on the assumption that *the formal description is a model of the physical reality.* When this is not the case – for example, if there is a power failure – no conclusions can be made from the formal considerations.

1.3.2 Simulation

It is possible to transform the formal description of a design into a program which can be executed on a computer to simulate the behavior of the design. This corresponds to building a prototype, and it makes it possible to experiment with the design by executing it on various sets of input data. Prototypes have always played an important role in circuit design; however, these are usually physical circuits that take a lot of effort to construct. In contrast, a formal description of a design – for example, in Synchronized Transitions – can automatically be turned into an executable program in a few seconds. It is, therefore, realistic to experiment extensively with various modifications of the design before conceiving low-level circuit details.

Consider again the traffic light controller described in section 1.3; below is shown a printout of the dialogue with an operator simulating this design. The operator provides the values of the state variables *CarOnEW*, *CarOnNS*, and the program then simulates the four transitions and outputs the value of the two state variables controlling the lights. The input provided by the operator is underlined.

```
CarOnNS: true    CarOnEW: false
EW = red NS = green
CarOnNS: false   CarOnEW: true
EW = green NS = red
CarOnNS: true    CarOnEW: false
EW = red NS = green
CarOnNS: true    CarOnEW: false
EW = red NS = green
```

For more complicated and interesting designs, simulation is a very important part of the design process, in particular to verify aspects that cannot be stated formally.

1.3.3 Tools

Formal methods need support by tools, just as design at other levels is supported by tools, for example, CAD tools for layout. Such high-level tools are still in their infancy; however, there are several prototype tools supporting circuit design using Synchronized Transitions. These tools are translators that transform a design description into various other forms. There are three main classes of such translators, helping the designer at different phases of the design:

simulation: produces an executable program in a programming language,

verification: produces verification conditions that can be checked by a mechanical theorem prover,

synthesis: produces circuit descriptions, e.g., net-lists or layouts.

Exactly the *same* source program can be fed to all three classes of translators. This prevents errors from creeping into a design when it is manually transformed from one form into another.

1.3.4 Terminology

This chapter has introduced some terminology that is used throughout this book. The most important terms are therefore repeated. Synchronized Transitions has many similarities with a programming language, however, there are also important differences. Therefore, the term *design* is used rather than program about the object being described. Sometimes it is necessary to refer to the textual description in Synchronized Transitions, which is then called a *design description*. A design description provides an *abstract model* of the design, and a *computation* is the dynamic behavior of a design.

In the text of this book, complete design descriptions are framed and appear as figures, whereas incomplete design fragments appear as mathematical formulae in the text. These are often incomplete and might contain symbols which are not valid in Synchronized Transitions, for example, $\ldots$ or 2^n.

1.4 BACKGROUND

The idea of using programming languages to describe hardware structures is not new, but for a long time such descriptions where only abstract models of the hardware with no possibility of automatically compiling a program into a realization. However, design tools for large-scale integration have made it feasible to automatically synthesize masks for circuit fabrication from symbolic descriptions. The first such tools were called "silicon compilers" [1, 15, 21]. Currently, the more modest term synthesis tools is the most common.

Circuit structures now appear as examples in papers on concurrent programming languages, for example, in the first published description of CSP [35]. More recently, CSP (or variations of it) has been used in several tools for hardware design [47, 49].

This book focuses on models based on concurrency; it has also been suggested to use functions and relations to model the computations of a circuit [24, 29]. This leads to models which are very different from the ones shown here, but most of the goals are the same, for example, to encourage the use of abstract models and to facilitate formal manipulations.

2

DESIGNING WITH TRANSITIONS

This chapter shows how to write design descriptions in Synchronized Transitions. The fundamental concepts of the language are introduced through a series of examples which also illustrates an asynchronous design style used throughout the book. Asynchronous behavior is a powerful abstraction where functionality is stressed, leaving timing details for later consideration. Such asynchronous designs can be realized in many different ways including synchronous circuitry, as shown in chapter 5, self-timed circuits, as shown in chapter 7, and software.

Chapters 3 and 4 introduce the remaining constructs of the Synchronized Transitions language, and appendix A contains a detailed language description including a formal definition of the syntax.

2.1 COMPUTATIONAL MODEL

Synchronized Transitions is a language with constructs for writing succinct, structured, and parameterized descriptions of large transition systems. Such a transition system consists of a fixed set of transitions operating in a state space.

The transition system is determined by a set of state variables, $s_1, s_2, \ldots, s_n$, and a number of transitions, $t_1, t_2, \ldots, t_m$. For a given design, the set of state variables and transitions are fixed. There is a fixed domain associated with each state variable determining the set of values that it can take. A state is a mapping of state variables to values, $(s_1 \mapsto v_1, s_2 \mapsto v_2, \ldots, s_n \mapsto v_n)$, where $v_i \;\; (1 \leq i \leq n)$ is a value in the domain of state variable s_i. A transition is a set of ordered pairs, for each such pair the first element is called the pre-state and the second the post-state.

A design defines a set of computations that are sequences of states, $S_0, S_1, \ldots$, where S_0 is an initial state, and for each pair, S_i, S_{i+1}, there is a transition, t, such that S_i is a pre-state of t, and S_{i+1} is a post-state of t. Note, that there is only a single transition accounting for one step of the computation (going from S_i to S_{i+1}).

Example: A Simple Oscillator

Let s_0, s_1, and s_2 be three boolean state variables. Hence, there are eight possible states. Below, three transitions are described that operate on the three state variables.

$$\ll s_0 \neq s_2 \texttt{->} s_1 := s_0 \gg \;||$$
$$\ll s_1 \neq s_0 \texttt{->} s_2 := s_1 \gg \;||$$
$$\ll s_2 \neq s_1 \texttt{->} s_0 := s_2 \gg$$

The first transition description defines two possible state changes, one leading from a state where s_0 is true and s_1, s_2 are false, and another where s_0 is false and s_1, s_2 are true. If this transition system is initialized such that the values of the three state variables are *not* the same, then it performs an infinite computation where each of the three state variables oscillates between true and false.

2.2 STATES

The state space of a transition system is determined by a set of state variables, each of which contains a value. These values define the current state, which is changed by assigning new values to one or more state variables. State variables are introduced by a variable declaration.

$$\mathit{STATE}\ \ CarOnNS : \mathit{BOOLEAN}$$

The name of the state variable (appearing before the ":") denotes the value of the variable. The type of the state variable (given after the ":") specifies its domain, i.e., the set of possible values. The state variable *CarOnNS* can, for example, take the boolean values *TRUE* and *FALSE*. The following shorthand is used for introducing two or more state variables of the same type.

$$\mathit{STATE}\ \ CarOnNS, CarOnEW, reset : \mathit{BOOLEAN}$$

The type of a variable determines the operations that can be applied to it. In case of booleans, it is the logical operators such as negation, logical OR, etc.

The value of a state variable is changed by executing a transition where the name of the state variable (e.g., *CarOnNS*) appears on the left-hand side of an assignment (:=). When the name of a state variable appears in an expression it denotes the most recent value assigned to it.

The set of state variables of a design is fixed and cannot be changed dynamically. Section 2.8 describes how to partition the state space into components with local state variables that are hidden from other components.

2.2.1 Initial State

The initial state is specified by defining the initial value of some or all of the state variables.

INITIALLY $CarOnNS = FALSE \;\; CarOnEW = FALSE$

State variables get their initial value before any transitions are executed, and they retain this value until a different value is assigned to them. It is not required that state variables are given an initial value.

2.2.2 Static Declaration

A static is a named quantity that does not change during the computation. Statics are used to give symbolic (readable) names for some values, for example, the integer values representing the colors of a traffic light. Statics are introduced by a static declaration.

STATIC $green = 0 \;\; yellow = 1 \;\; red = 2$

Statics are used in expressions just like names of state variables, thus the name *green* can be used instead of writing *0*.

2.3 TRANSITIONS

The state changes of a transition system are specified by a set of transition descriptions, for example:

$\ll$ *NSlight* $=$ *yellow* $->$ *EWlight* $:=$ *red* $\gg$

This describes a transition changing the state by assigning the value *red* to the state variable *EWlight*. The state change is only made if the precondition *NSlight* $=$ *yellow* holds. The precondition must be a boolean expression. A transition is enabled in a particular state, if its precondition holds in that state. The state change is called the execution of a transition.

In general, the change of state variables is described by a multi-assignment, where several state variables are given new values.

$\ll$ *NSlight* $=$ *yellow* $->$ *EWlight*, *NSlight* $:=$ *red*, *green* $\gg$

In a multi-assignment, all expressions on the right-hand side are evaluated and then assigned to the variables on the left-hand side. All variables on the left-hand side must be distinct.

When the precondition is the constant *TRUE*, it can be omitted. For example, the transition description:

$\ll y := a\ OR\ b \gg$

is equivalent to:

$\ll TRUE\ ->\ y := a\ OR\ b \gg$

The set of transitions specified by a design is fixed and cannot be changed dynamically. Section 2.8 describes how the set of transitions can be partitioned into components with local transitions operating on local variables.

2.3.1 Operational Model

A design description specifies a set of transitions. This set is fixed throughout the computation, and *all* transitions may execute whenever they are enabled. Although a design description in Synchronized Transitions has some similarity with a program in a high-level programming language, the interpretation is very different. An assignment statement in a high-level program is only executed when the control of the program is at the point of the statement. There is no similar global control flow determining the computations of a design in Synchronized Transitions, which just specifies a fixed set of transitions. Unlike a Pascal or C program, the order in which transition descriptions are written does not influence the computation.

Operationally, the computation can be modelled as repeated nondeterministic selection and execution of an enabled transition. In this model, a transition is executed:

- *repeatedly,* each time it has been executed, it is immediately ready to be selected again,
- *atomically,* evaluation of the precondition and performing the multi-assignment is an indivisible operation,
- *independently,* of the order it appears in the design description.

In this model, it is not required that a transition is executed immediately after it becomes enabled, because other enabled transitions may be selected. In fact, there is no upper bound on when a transition is selected. This is an abstraction of the delays caused by the components in physical circuits. For example:

$$\ll y := a \ OR \ b \gg$$

describes an OR-gate. The implicit precondition, $TRUE$, specifies that it is always allowed to set the output, y, to the logical OR of the inputs, a and b; however, an arbitrary delay may elapse between a change of the inputs and the changing of the output. The pre-state is the state immediately before executing the transition and it leads to the post-state.

This is a very concrete and simple model of a computation, and it must be stressed that it is only a model, and a realization may execute transitions differently, as long as the realization does not enter states that cannot be entered by the model. As explained in sections 5.4.1 and 7.3.1, it is often possible to ensure atomicity, even if several transitions are executed in parallel, thus allowing efficient circuit realizations where many subcircuits operate simultaneously. One possibility is to implement transitions as subcircuits and state variables as connections between the subcircuits. This is discussed in further detail in section 5.2.

The asynchronous model described in this chapter is not the only way of associating a computation with a set of transitions. Chapter 4 introduces another computational model used for synchronous computations. In order to distinguish these different computational models, the asynchronous combinator, ||, is used for describing asynchronous designs.

```
STATIC green = 0 yellow = 1 red = 2
STATE
  EWlight, NSlight: INTEGER
  CarOnEW, CarOnNS: BOOLEAN
INITIALLY
  EWlight = green   NSlight = red
  CarOnNS = FALSE   CarOnEW = FALSE
BEGIN
  ≪ (EWlight = green) AND CarOnNS -> EWlight:= yellow ≫ ||
  ≪ (EWlight = yellow) -> EWlight, NSlight:= red, green ≫ ||
  ≪ (NSlight = green) AND CarOnEW -> NSlight:= yellow ≫ ||
  ≪ (NSlight = yellow) -> NSlight, EWlight:= red, green ≫
END
```

Figure 2.1 A simple traffic light controller

Terminology

The write set, W^t, of a transition description, t, is the set of state variables appearing on the left-hand side of assignments. Similarly, the read set, R^t, of a transition description is the set of state variables that appear in the precondition and on the right-hand side of the assignment (see appendix A section A.14.1). A transition is enabled if the precondition holds.

Example: A Traffic Light Controller

Figure 2.1 shows the description of a traffic light controller (discussed in section 1.3) written in Synchronized Transitions. The design description shown in figure 2.1 defines a transition system with four transitions and four state variables, two of which are integers and two of which are boolean. The three statics *green*, *yellow*, and *red* denote the values *0*, *1*, and *2* respectively.

In this particular design, at most one transition is enabled at a time, for example, in a state where $NSlight = green$ and there is a car on the East-West road so that $CarOnEW$ is true, the third transition can execute. This changes $NSlight$ to $yellow$, enabling the fourth transition, etc. It is not described how $CarOnEW$ and $CarOnNS$ change when cars are approaching the intersection.

2.4 ARRAYS AND QUANTIFICATION

Circuits where the same elements are replicated many times are quite common. By using arrays and quantification, it is possible to describe a number of similar state variables and transitions in a succinct way. An array type defines a vector of values of the same type.

$$TYPE\ word = ARRAY\,[0..n-1]\ OF\ BOOLEAN$$

A value of type word is a vector of n booleans. State variables of type word take such values, for example:

$$STATE\ a, b, s :\ word$$

The i'th element, $0 \leq i < n$, of the state variable a is denoted $a[i]$, the type of this element is boolean. An element of an array, e.g., $a[i]$ is used the same way as any other state variable, it may for example occur in expressions like:

$$NOT\ a[i]\ AND\ CarOnNS$$

One may describe a computation where a number of similar transitions operate on different array elements as follows:

$$\{\ \|\ i :\ [0..n-1]\ |\ \ll s[i] := a[i]\ OR\ b[i] \gg \}$$

This is called a quantified instantiation, and it describes a set of n different transitions, one for each i in the range $0..n-1$. Hence, it is a compact way of writing the following:

$$\ll s[0] := a[0]\ OR\ b[0] \gg \ \|$$
$$\ll s[1] := a[1]\ OR\ b[1] \gg \ \|$$
$$\ldots$$
$$\ll s[n-1] := a[n-1]\ OR\ b[n-1] \gg$$

These two descriptions specify exactly the same sets of transitions and hence the same computation. However, the second description can be cumbersome to write when n is large. Furthermore, the quantified instantiation allows the designer to write a generic description where the same textual description is used for different values of n. It is, however, important to stress that whether or not quantified instantiation is used to describe the set of transitions, the set is *fixed* throughout the computation. As explained in section 2.3.1, any transition may execute whenever it is enabled. There is no global control flow "calling" transitions as is the case in a programming language.

Example: Ripple Carry Adder

Using the constructs introduced so far, it is possible to describe a generic n-bit ripple carry adder with the same functionality as the one introduced in section 1.3.

$$\{\| \; i: \; [0..n-1] \mid \ll s[i] := (a[i] \neq b[i]) \neq c[i] \gg \} \; \|$$
$$\{\| \; i: \; [0..n-1] \mid \ll c[i+1] := (a[i] \; AND \; b[i]) \; OR \; (a[i] \; AND \; c[i]) \; OR \; (c[i] \; AND \; b[i]) \gg \} \; \|$$
$$\ll c[0] := FALSE \gg$$

This describes $n+n+1$ transitions. The first n show how the bits of the sum are computed. The next $n+1$ how the carry is obtained. It is important to note that this describes a computation where $2n+1$ transitions are executing concurrently and repeatedly, just like the $2n+1$ subcircuits they model. Therefore, the order in which new values appear in $s[i]$ and $c[i]$ cannot be predicted. So, a particular $s[i]$ may change value many times before it stabilizes at the correct value. This stabilization progresses from the least significant bit (low index) to the most significant. Exactly the same happens in the circuit. It may seem odd that the circuit computes these spurious values before stabilizing. It could be avoided by describing the computation sequentially, but then it would no longer be an adequate abstraction of the circuit.

2.5 FIXED POINTS

The ripple carry adder illustrates an important concept called a fixed point that is a state where no further state changes are possible. At a fixed point all transitions are either disabled or executing them does not assign new values to state variables. If a design is completely isolated it will remain in the fixed point state forever.

Example: Greatest Common Divisor

In figure 2.2 is shown a design for computing the greatest common divisor of two integers, X and Y. When $X > Y$, it is known that:

$$gcd(X, Y) = gcd(X - Y, Y)$$

and when $Y > X$ that

$$gcd(X, Y) = gcd(X, Y - X)$$

```
STATE x,y: INTEGER
INITIALLY x=X  y=Y
BEGIN
  ≪ x > y -> x:= x-y ≫  ||  ≪ x < y -> y:= y-x ≫
END
```

Figure 2.2 Computation of the greatest common divisor

These simple observations are the basis of the algorithm shown in figure 2.2. The initial values of the state variables x and y are X and Y respectively. The computation reaches a fixed point where $x = y$. In this case, it is simple to verify that this is a fixed point, since none of the transitions are enabled when $x = y$. Furthermore, when $x = y$ then $x = gcd(x, y)$, and since $gcd(x, y) = gcd(X, Y)$ throughout the computation, it follows that when the computation is at a fixed point then:

$$x = gcd(X, Y)$$

Hence, the greatest common divisor of X and Y is the value of x (or y, which has the same value).

2.6 STATICS

In section 2.2.2, statics were introduced to give symbolic (readable) names for certain values. Statics can also be used for specifying bounds, for example, the number of bits in a register.

STATIC n = *16*
TYPE
  $range = [0..n-1]$
  $word$ = *ARRAY* $range$ *OF BOOLEAN*

The number of bits in a register or an adder cannot be changed dynamically, and therefore the bounds are specified with static expressions. Statics are also introduced in quantified instantiations, for example:

$\{\| \; i : range \mid \ll s[i] := a[i] \; OR \; b[i] \gg \}$

The range variable, i, is a static, because in each particular transition it denotes a fixed value.

	a	b	c	d	e	f	g
a		2	8	10			
b			3				
c						4	
d			12			5	
e						16	
f	2	4	12	3			1
g		2			12		

Figure 2.3 Example of a cost matrix

Example: All Pairs Shortest Paths

This section presents a solution to an optimization problem, usually known as the all pairs shortest paths. The solution is an example of a high-level design, using data types and operations that are not directly available at the circuit level. In chapter 6 techniques are described for refining such a high-level design into a realization. However, in the early phases of a design, it can be useful to work at a high level without considering the details of a realization.

Consider a table of prices for flying between various cities (a cost matrix), called $a, b, c, \ldots$ (see figure 2.3).

```
TYPE
  range = [1..n]
  row = ARRAY range OF INTEGER
  table = ARRAY range OF row
STATE
  d : table
```

The cost matrix shows that the price of flying from a to c is 8, and when flying from e to f, it is 16, etc. The table does not give any price of flying from a to f; however, by going through d the price becomes 15. The question is, whether this is the cheapest way of getting from a to f? Maybe, it can be done cheaper by going through d, e, or by making more than one stop-over. To answer this question, we need an algorithm that can compute the cheapest way of getting between any pair of cities. Absence of an entry in the cost matrix, W, indicates that there is no direct connection between the corresponding two cities. The cost is assumed to be infinite (or some finite approximation that is larger than all other costs).

The problem is to compute the cheapest way of getting between any pair of cities allowing routes going through an arbitrary number of intermediate cities.

Assume that $d[i][j]$ contains an estimate of the cheapest way of getting between i and j. If there is a city k such that $d[i][k] + d[k][j] < d[i][j]$, it means that a cheaper route between i and j has been found, and hence $d[i][j]$ should be updated. This is expressed with the transition.

$$\ll d[i][k] + d[k][j] < d[i][j] -> d[i][j] := d[i][k] + d[k][j] \gg$$

This transition is executed repeatedly, whenever:

$$d[i][k] + d[k][j] < d[i][j]$$

the improved estimate, is assigned to $d[i][j]$. Similar transitions update $d[i][j]$ for all other $n-1$ possible intermediate cities, leading to a design where there are n transitions for each pair of cities. All the transitions are of the same form as the one shown above, but each is working on its own triple of indices. This is described as follows:

$$\{|| i, j, k : [1..n] \mid \ll d[i][k] + d[k][j] < d[i][j] -> d[i][j] := d[i][k] + d[k][j] \gg\}$$

This describes n^3 transitions that work concurrently and repeatedly. Every time $d[i][j]$ is updated, a cheaper route has been found, which may lead to improvements in other routes, etc. All costs in the initial cost matrix, W, are assumed to be positive (otherwise one could earn money by travelling certain routes), which means that sooner or later the computations reaches a fixed point.

$$\forall i, j, k : \mathit{range} \,.\; d[i][k] + d[k][j] \geq d[i][j]$$

When this assertion holds, $d[i][j]$ contains the minimal cost of getting between cities i and j. This is shown below by assuming it is not the case and deriving a contradiction. It is important to note that *throughout the computation* there is always a route connecting i and j with the cost $d[i][j]$. This is also the case at the fixed point.

The design description is used to argue that the algorithm computes the cheapest route. This can be seen by assuming the converse and then deriving a contradiction. Therefore, assume that $d[i][j]$ is not the minimal cost of getting between the cities i and j, then there exists one or more routes between i, j with a cost less than $d[i][j]$. Pick a route, p, connecting a pair of cities i, j, i.e.,

$$p : i = v_0, v_1, v_2, \ldots, v_l = j$$

with the following two properties:

1. $c_p < d[i][j]$,

2. there is not a route with fewer than l stops for which d does not represent the minimal cost.

If there are two or more routes with the same number of stops satisfying these two properties, one of them is picked arbitrarily. The cost c_p is:

$$c_p = d[i][v_1] + d[v_1][v_2] + \ldots + d[v_{l-1}][j]$$

Because there is no route with fewer stops than p for which $c_p < d[i][j]$, it must be the case that

$$d[v_1][v_2] + \ldots + d[v_{l-1}][j] \geq d[v_1][j]$$

Therefore,

$$c_p \geq d[i][v_1] + d[v_1][j] \geq d[i][j]$$

This is a contradiction with the assumption $c_p < d[i][j]$, which therefore cannot hold. Hence, it has been shown that for any pair of cities i, j, the cheapest cost of getting between them is the value of $d[i][j]$ at the fixed point.

This design for finding the cheapest route between all pairs of cities has some similarities with the adder presented in section 2.4. Both can be characterized as a chaotic computation where a number of similar and simple transitions collectively work towards a common goal. The concurrent execution is what makes the transitions reach the goal. As pointed out above, the all pairs shortest path algorithm is a high-level design where many details that are important for an efficient realization are left open. In the case of the adder, it is not so difficult to see the relationship between the design description and a reasonably efficient circuit, but for the all pairs shortest path computation, it is not so clear. On the other hand, the high-level description of the all pairs shortest path algorithm provides an abstract model where it is possible to verify the functional correctness of the algorithm which is not immediately obvious. It would be very cumbersome to verify this at the circuit level. In this example, the model of the computation described in Synchronized Transitions provides an adequate abstraction for the functional verification.

Example: Carry Lookahead Adder

This section describes the design of a carry lookahead adder which is a faster adder than the simple ripple carry adder described in section 2.4. Furthermore, it illustrates the use of quantification to describe a replicated structure.

The speed of the simple ripple carry adder is limited by the time it takes to propagate the carry, which in the worst case is proportional to the number of bits in the operands (n). To speed up the addition, one may use a carry lookahead adder. Such an adder generates all carry bits in parallel, instead of doing it sequentially as in the ripple carry adder. Assume for a moment that all the correct carry bits have been determined; then all the sum bits can be computed by the following transitions:

$$\{\| \; i : \; [0..n-1] \mid \ll s[i] := (a[i] \neq b[i]) \neq c[i] \gg \}$$

Because $s[i]$ does not depend on any other bits of the sum (elements of s), all bits, $s[i]$, can be computed in parallel. The careful reader might question the use of the word parallel here, because in section 2.1, it is explicitly stated that transitions are executed atomically. If atomicity is realized by executing transitions one at a time then the sum bits are certainly not computed in parallel. However, in many cases atomicity can be ensured even if transitions are executed simultaneously. The adder is an example of this, because none of the state variables that are changed ($s[i]$) are read by the other transitions. This is explained in further detail in section 5.4.1.

To compute the sum, the carry bits ($c[i]$) must be determined. These are also computed in parallel by introducing two auxiliary variables called g (generate) and p (propagate).

$$\ll g[i] := a[i] \; AND \; b[i] \gg \; \| \; \ll p[i] := a[i] \; OR \; b[i] \gg$$

The i'th carry bit is computed from $p[i], g[i]$ as follows:

$$c[i+1] = g[i] \; OR \; (p[i] \; AND \; c[i])$$

The expression for $c[i+1]$ includes $c[i]$, but this can be eliminated by substituting the expression for $c[i]$ into the expression for $c[i+1]$. This yields:

$$c[i+1] = g[i] \; OR \; (p[i] \; AND \; (g[i-1] \; OR \; (p[i-1] \; AND \; c[i-1])))$$

By repeating the substitution i times, one obtains an expression for $c[i+1]$ using only p's and g's, which in turn are expressed directly from the input operands, a and b.

$$c[i+1] = g[i] \; OR \; (p[i] \; AND \; (g[i-1] \; OR \; (p[i-1] \; AND \; (g[i-2] \; OR \; ...))))$$

Hence, an expression is obtained where all the carry bits are computed in parallel. But what about the "..." in the expression shown above? It can be eliminated by using a quantified expression. Syntactically, it is similar to a quantified instantiation. In general, it is written as follows:

$$\{ \; OR \; i : \; [0..n-1] \mid g[i] \; \}$$

$\ll$ c[0]:= FALSE $\gg$ ||
{ || i: [0..n-1] | $\ll$ g[i]:= a[i] AND b[i] $\gg$ || $\ll$ p[i]:= a[i] OR b[i] $\gg$ ||
$\ll$ c[i+1]:= { OR j: [0..i] | g[i-j] AND { AND k: [0..j-1] | p[i-k] }} $\gg$ ||
$\ll$ s[i]:= (a[i] $\neq$ b[i]) $\neq$ c[i] $\gg$ }

Figure 2.4 Transitions describing a carry lookahead adder

that is a succinct way of writing the expression

$$g[0]\ OR\ g[1]\ OR\ g[2] \ldots g[n-1]$$

If the range of the quantification is empty (the upper limit is smaller than the lower limit) then the value of the quantified expression is the neutral element of the operator (*FALSE* in case of OR and *TRUE* in case of AND).

Two nested quantified expressions are used to describe the carry computation in the carry lookahead adder.

$$\ll c[i+1] := \{\ OR\ j : [0..i] \mid g[i-j]\ AND\ \{\ AND\ k : [0..j-1] | p[i-k]\}\} \gg$$

This may look a bit complex at first sight; however, two important things are achieved. Firstly, the ... is avoided, and secondly one obtains a general expression that can be used for all values of n. The carry transitions of a three bit adder can also be written as follows:

$$c[1] := g[0]\ ||$$
$$c[2] := g[1]\ OR\ (p[1]\ AND\ g[0])\ ||$$
$$c[3] := g[2]\ OR\ (p[2]\ AND\ (g[1]\ OR\ (p[1]\ AND\ g[0])))$$

However, to specify a four bit adder, an additional transition is required. Hence, a different description is required for each value of n. By using a quantified expression a single description is used for all values of n. The complete carry lookahead adder is shown in figure 2.4.

This adder appears to be very fast, because all the carry propagate and generate values are computed in parallel with each other and in parallel with computing all the carry and sum bits. However, a closer look at the carry lookahead adder reveals a computation of a logical AND/OR on n inputs. For large values of n, it is unrealistic to assume that an AND operation on n inputs is computed in a small constant time. The time needed must necessarily grow with n. Furthermore, when the design is realized as a circuit, such operations are not directly available, they must be constructed out of gates with a limited number of inputs, typically 2, 3, or 4. In section 2.9, it is shown how an n-input AND-gate can be constructed out of 3-input AND-gates. By organizing the computation in a tree, it is possible to obtain a computing time proportional to $log\ n$.

2.7 NAMED TRANSITIONS

A named transition specifies a generic transition that can be instantiated one or more times. This mechanism does not provide any additional expressive power; however, it can make a design more comprehensible. A parameter mechanism allows each transition to operate on different state variables. This is an example of a named transition:

TRANSITION $propagate(x, y, z$: *BOOLEAN*)

$\ll z := x\ OR\ y \gg$

There are three formal parameters x, y, and z; these are like state variables, they can appear in expressions and on the left-hand side of assignments. Formal parameters have a type, just like state variables; in this case it is $BOOLEAN$. The named transition is a template for one or more similar transitions, for example:

$propagate(a[1], b[1], p[1])\ ||\ propagate(a[2], b[2], p[2])$

Here two transitions are instantiated, one which operates on the state variables $a[1], b[1], p[1]$, and one which operates on $a[2], b[2], p[2]$. The state variables appearing in the instantiation are called actual parameters. In the first transition all occurrences of the formal parameter x are substituted by $a[1]$, and all occurrences of y are substituted by $b[1]$, etc. Hence, the two instantiations shown above yields the following two transition descriptions:

$\ll p[1] := a[1]\ OR\ b[1] \gg\ ||\ \ll p[2] := a[2]\ OR\ b[2] \gg$

The actual parameters must be state variables. It is not allowed to pass an expression, e.g., $d + e$ as an actual parameter, because such an expression denotes a value and not a state variable.

Example: A FIFO Queue

This example illustrates the use of named transitions. It shows how to design a FIFO queue, and it is also an example of a design at a somewhat higher level than the adder. A FIFO queue is a data structure that holds a sequence of elements. There are two external operations on a FIFO queue, insertion and

```
TYPE vector = ARRAY [0..n-1] OF element
STATE
   q: vector
   in, out: element
INITIALLY {i: [0..n-1] | q[i] = E }
TRANSITION transfer(from, to: element)
   ≪ from ≠ E  AND  to=E  ->  to, from:= from, E ≫
BEGIN
   transfer(in, q[0]) || transfer(q[n-1], out) ||
   {|| i: [0..n-2] | transfer(q[i], q[i+1]) }
END
```

Figure 2.5 FIFO design

removal. Elements are always removed in the same order as they are inserted (First In First Out).

The interface to the FIFO consists of two registers, an input register *in*, and an output register, *out*.

The static E (empty) denotes a value that is different from all values contained in the queue. An element can be inserted in the queue when the input register is empty, i.e., it has the value E. Insertion is done by assigning an element to *in*. An element can be removed when the output register is full ($out \neq E$). The removal is done by changing the value of *out* to E. If the queue is not empty, the output register *out* then gets the value of the oldest element which is presently in the queue.

The FIFO queue is constructed from an array of registers (each of which can hold an element), and a number of transitions for copying elements (see figure 2.5). The type *element* must contain the distinct value E in addition to all the proper values of elements. Initially, all registers in the queue are empty. When an element is inserted through *in* it moves down the queue towards *out*.

```
CELL fifo(in, out: element)
   TYPE vector = ARRAY[0..n-1] OF element
   STATE q: vector
   INITIALLY {i: [0..n-1] | q[i] = E }
   TRANSITION transfer(from, to: element)
      ≪ from ≠ E  AND  to=E –> to, from:= from, E ≫
BEGIN
   transfer(in, q[0])  ||  transfer(q[n-1], out) ||
   {|| i: [0..n-2]  |  transfer(q[i], q[i+1]) }
END fifo
```

Figure 2.6 FIFO cell

Meanwhile, further elements can be inserted. Several elements can move in parallel. In this example, in, out is the external interface to the queue whereas q is an internal variable that could be hidden from the outside. In the next section, the cell concept is presented; this makes it possible to encapsulate part of a design, for example, by hiding internal variables such as q.

2.8 CELLS

It is rarely feasible to view a nontrivial design as a single monolithic transition system. Such a design becomes easier to manage if it is broken into a number of (almost) independent pieces that can be considered separately. This section presents the cell concept that is used to modularize designs.

A cell encapsulates part of a design, and it is a generic (i.e., parameterized) description of a set of state variables and transitions called the local state variables and local transitions of the cell. Formal parameters define the interface of a cell. The FIFO queue presented in section 2.7 can be described as a cell with an interface consisting of two formal parameters in, out and with a local state variable q (see figure 2.6). The internal structure of a cell, for example the variable q, is hidden from the rest of the design. The scope of local state variables (and formal parameters) is the cell in which they are declared.

2.8.1 Cell Instances

A cell describes a parameterized part of a design. An instance of such a part is described by a cell instantiation consisting of the cell name and a list of actual parameters.

$$\mathit{fifo}(ireg, oreg)$$

This describes a set of transitions operating on the state variables $ireg$, $oreg$ and a local copy of q. Another copy of the fifo can be instantiated.

$$\mathit{fifo}(a, b)$$

This cell instance operates on a, b and a new local copy of q.

Communication between two (or more) different cell instances is possible if the same actual parameter is passed to each of them, for example:

$$\mathit{fifo}(a, b) \;||\; \mathit{fifo}(b, c)$$

The actual parameter b is used by the first instance as an output register and by the second instance as an input register. It is also possible for other transitions to use b, for example:

$$\mathit{fifo}(a, b) \;||\; \mathit{fifo}(b, c) \;||\ll b{\neq}E \mathrel{->} disp := b \gg$$

This illustrates how an actual parameter can be used by both the instantiating and the instantiated cell. Section 2.10 shows how to restrict where a state variables can be changed.

The sets of transitions of different cell instances are disjoint, hence, any transition belongs to exactly one cell instance. It is important to stress that a cell instantiation specifies a statically fixed set of transitions. As it has been mentioned several times already there is no global control flow that dynamically creates or removes transitions. Hence, a cell instantiation such as $\mathit{fifo}(a, b)$ is very different from a procedure call in a programming language that dynamically allocates variables, etc. Figure 2.7 illustrates the relationship between a design description and a computation. It is best illustrated in two steps, the first, called the instantiation, consists of generating a transition system from a design description. This results in a static set of transitions. In the second step, called the execution, state changes are made as described by transitions in the transition set. During execution the set of transitions cannot change. The collection of all cell instances in a modular design defines a transition system, where the state space is the union of the state spaces of the cell instances, and the set of transitions is the union of the transitions of the cell instances. The computation is a sequence of states in this state space, corresponding to atomic

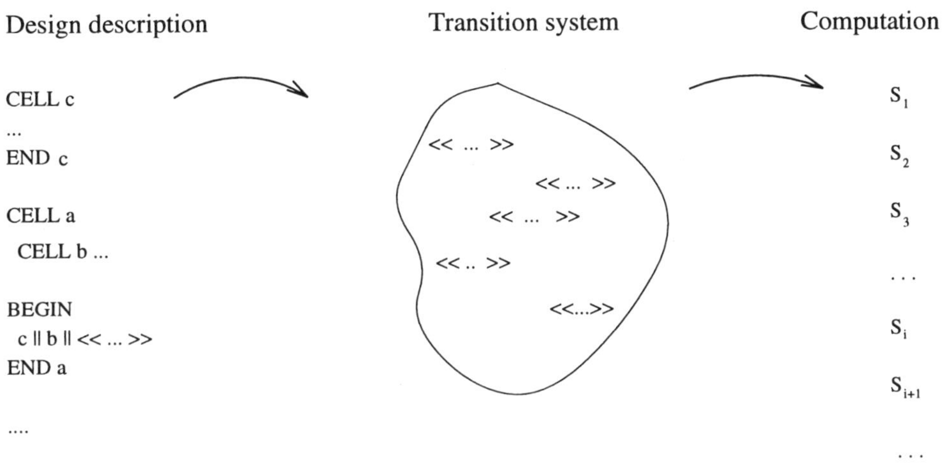

Figure 2.7 Relation between design description and computation

executions of the transitions.

Example: An Adder Cell

Figure 2.8 shows the design of a ripple carry adder, and it illustrates how cells can be nested. The cell *adder* has a local cell, *full_adder*, and a local variable c. The body of the adder, instantiates n copies of the *full_adder*.

2.8.2 Instance Tree

The cell instantiations of a design form a tree of cell instances, called the instance tree. The root of the tree is an instance of a designated main cell of the design. There is an edge from the root to a number of cell instances, one for each cell instantiation given in the main cell.

An instantiation of a cell B from a cell A in the design corresponds to a number of edges (a, b) in the tree, where a is an instance of A and b is an instance of B. The nodes a and b are said to be neighbors in the tree, and b is said to be a subcell instance of a. This is illustrated in figure 2.9.

A cell instance, e.g., d, is said to be interior to another cell instance, c, if d is below c in the instance tree, i.e., if c is on the path from the root to d (and they

```
TYPE
   range = [0..n-1]
   word = ARRAY range OF BOOLEAN
CELL adder(a, b, s: word)
   TYPE cword = ARRAY [0..n] OF BOOLEAN
   STATE c: cword
   CELL  full_adder(a, b, s, ci, cout: BOOLEAN)
   BEGIN
      ≪ s:= (a ≠ b) ≠ ci ≫ ||
      ≪ cout:= (a  AND  b)  OR  (a  AND  ci)  OR  (ci  AND  b)≫
   END full_adder
BEGIN
   ≪ c[0]:= FALSE ≫ ||
   {|| i: [0..n-1] | full_adder(a[i], b[i], s[i], c[i], c[i+1]) }
END adder
```

Figure 2.8 Hierarchical design of a ripple carry adder

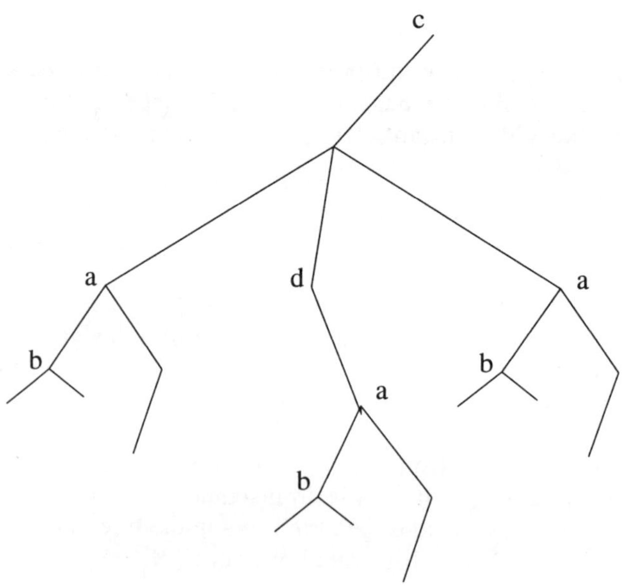

Figure 2.9 Example of an instance tree

are different nodes), see figure 2.9.

When there are several instances of a cell, the different instances can operate on different state variables (without this possibility all instances would do exactly the same). This is specified by allowing a cell to pass state variables as parameters. By giving different parameters in different cell instantiations, the corresponding cell instances effectively operate on different state variables.

2.8.3 Static Parameters

In addition to state variables, a cell may have static parameters. These are prefixed with the reserved symbol *STATIC*.

CELL component(reset : BOOLEAN; STATIC size : INTEGER)

Static parameters are used to control instantiations and to determine the size of data structures, e.g. arrays. The value of a static parameter is fixed when the cell is instantiated; therefore, the corresponding actual parameter must be an expression consisting exclusively of constant values and statics. Note that an actual static parameter can be an expression (without state variables).

Example: The Iterative Ring

An iterative ring is a cyclic computing structure consisting of a number of identical stages each capable of computing a function, F. Data is circulating in the ring and the function F describes the transformation made in each stage. Such an iterative ring can, for example, be used to implement a multiplier by choosing F as a shift and conditional accumulation.

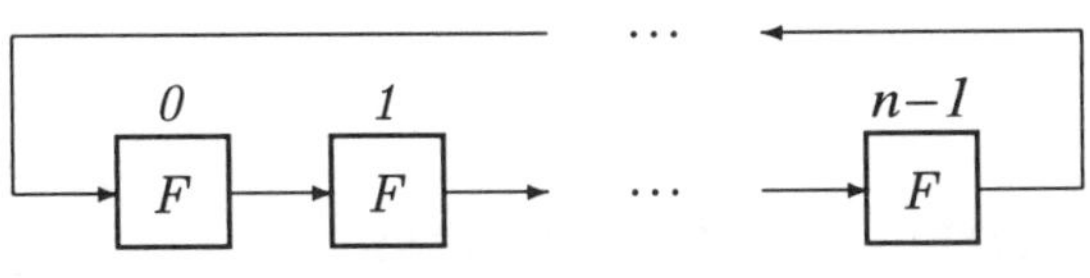

For simplicity, issues of input to and output from the ring are not considered here. Each stage of the ring is capable of computing F. The communication between the stages must ensure that the proper sequence of values is computed. In general, it is not possible to predict when one stage is ready to communicate with the next. If for example, F is a multiplier, the number of steps needed to propagate the carry (in the accumulator) varies. Therefore, the

```
CELL ring(STATIC n: INTEGER)
  STATIC E = 0 (* empty value *)
  TYPE
    link = INTEGER
    range = [0..n-1]
    vector = ARRAY range OF link
  STATE v: vector
  CELL stage(in, out, succ: link)
  BEGIN
    ≪ (in=E) ≠ (succ=E) -> out:= F(in) ≫
  END stage
BEGIN
  {|| k: range | stage(v[k], v[(k+1) MOD n], v[(k+2) MOD n]) }
END ring
```

Figure 2.10 Iterative ring

interface between the stages of the ring follows an asynchronous protocol. An asynchronous communication can be arranged in many different ways; in this particular case it is done by inserting spacers, called empty values, between consecutive nonspacer values, called valid values. The empty value is different from all valid values. This enables a stage to determine when a new valid value arrives, and hence when a new iteration of F can be computed. The same protocol is used in the FIFO queue presented in section 2.7. The empty value is denoted E. It is assumed that the function F maps the empty value to itself and a valid value to a valid value. Figure 2.10 shows the design of an iterative ring.

2.9 CONDITIONAL INSTANTIATION

Conditional instantiation is used to control whether or not a particular instantiation is done. A conditional instantiation is written as follows:

$$\{\ size > 0 \mid component(init,\ size-1)\ \}$$

The expression appearing before | is a boolean expression, and the instantiation following | is only done if the value of the boolean expression is true. The

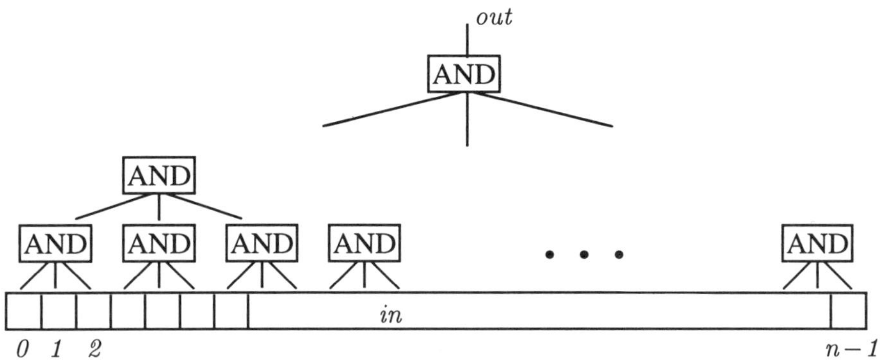

Figure 2.11 Tree of AND-gates

instantiation can be a transition description, a cell instantiation, a quantified instantiation or, another conditional instantiation. It must be possible to determine the value of the expression (in front of |) statically; therefore, it may only contain statics and constant values, but no state variables.

Conditional instantiation determines whether or not a particular subset of transitions is included in the transition system determined by a design, hence it directs the instantiation (see figure 2.7). This is in contrast to the precondition of a transition that determines whether a particular transition can execute, hence it directs the computation.

Example: Tree of AND-gates

As noted several times in this chapter, it is advantageous to design a VLSI circuit by replicating the same subcircuit many times. Quantification is one way of describing such replication, and in this section an alternative is shown using recursion to give a succinct description of a circuit consisting of many identical subcircuits. Recursion is a description mechanism which unfolds during instantiation and *not while the transitions execute.* To illustrate the use of recursive descriptions consider the n-input AND-gate that was needed in the carry lookahead adder (section 2.6). The problem is to compute the logical AND of n inputs. Assume that the inputs are placed in an array, in, of type word.

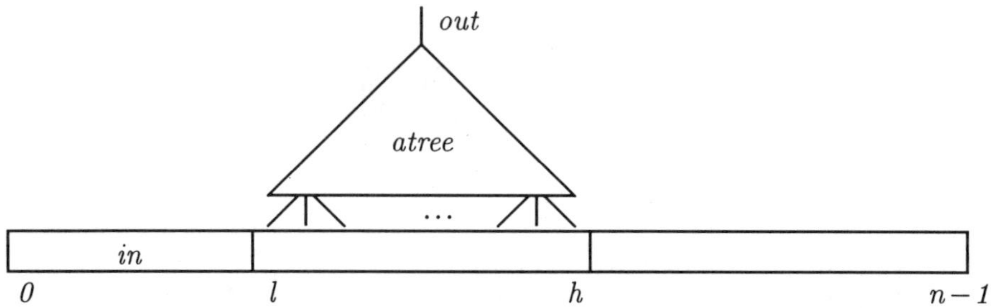

Figure 2.12 A partial tree

TYPE
$range = [0..n-1]$
$word$ = *ARRAY* $range$ *OF BOOLEAN*

Furthermore, assume that only 3-input AND-gates are available for realizing the computation. For simplicity assume also that n is a power of 3, i.e., for some integer $k > 0 : n = 3^k$. The n-input AND-gate can be constructed as a tree of 3-input AND-gates (see figure 2.11). It is now shown how such a tree is described with Synchronized Transitions.

The cell $atree(in, l, h, out)$ describes a tree of AND-gates covering the range from index l to index h of in, which gives the result in out (see figure 2.12). If such a cell is available, the complete tree is described as follows: the input array in is divided into three parts of the same size (it was assumed that n is a power of 3). Let w denote the size of each of these, i.e. $w = n/3$. For each of these parts an instance of the cell *atree* is made. These compute three partial results, namely the logical and of a $1/3$ of the input (see figure 2.13). In Synchronized Transitions this is described as follows:

$atree(in, 0, w-1, a1) \;||$
$atree(in, w, 2 * w-1, a2) \;||$
$atree(in, 2 * w, n-1, a3) \;||$
$\ll out := (a1 \; AND \; a2) \; AND \; a3 \gg$

The combination of these three trees is itself a tree, covering the input range from 0 to $n-1$. What has been achieved is a description of a tree as a combination of simpler trees, which could again be described as a combination of trees, etc. Such a recursion must stop somewhere, and in this case an *atree* with just three inputs cannot be subdivided further.

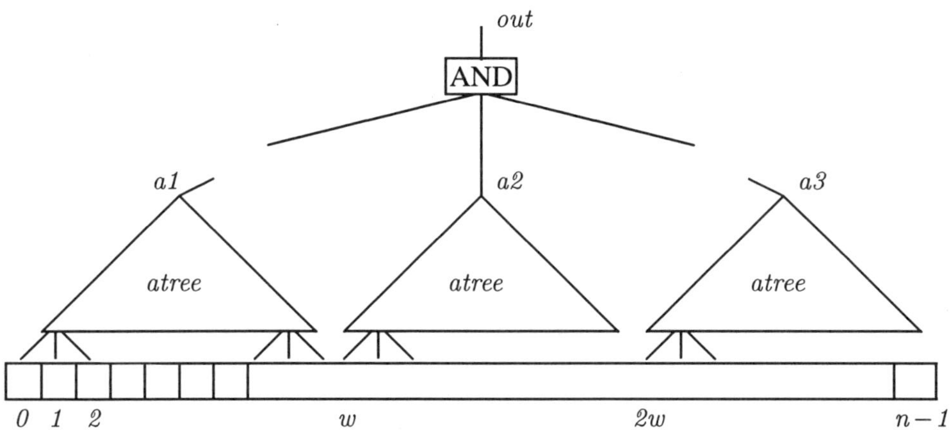

Figure 2.13 Recursive structure of an AND tree

Considering a general atree, $atree(in,\ l,\ h,\ out)$, there are two cases to consider: $h-l > 2$, and $h-l = 2$ (because n is a power of 3, $h-l$ cannot become 0 or 1). When $h-l = 2$ the *atree* is just a single 3-input AND-gate, when $h - l > 2$ it consists of three subtrees:

$$atree(in, l, l + w - 1, a1) \ \ ||$$
$$atree(in, l + w, l + 2 * w - 1, a2) \ \ ||$$
$$atree(in, l + 2 * w, h, a3)$$

where w is $(h–l+1)/3$. These two alternatives are described by two conditional instantiations, for example:

$$\{ \ h-l = 2 \mid \ll out := in[l] \ AND \ in[l + 1] \ AND \ in[l + 2] \gg \} \ \ ||$$
$$\{ \ h-l > 2 \mid \ll out := a1 \ AND \ a2 \ AND \ a3 \gg \ \ || \ldots \}$$

The recursive structure of the *atree* is utilized in the description by defining the cell *atree* recursively, as shown in figure 2.14. Instantiating the cell $atree(s, 0, 26, res)$ recursively instantiates 13 cells in a tree with 3 levels, 9 cells at the bottom level, 3 at the middle level and 1 root cell. Note that recursion is a description mechanism that unfolds during instantiation. Cells are used to structure a design description, and this structure controls the instantiation of transitions. Instantiations are made *statically*. Transitions, like hardware, are neither dynamically created nor destroyed.

```
TYPE
  range = [0..n-1]  (* assuming n = 3^k, k > 0 *)
  word = ARRAY range OF BOOLEAN
CELL  atree(in: word; STATIC l,h: INTEGER; out: BOOLEAN)
  STATIC w = (h-l+1) DIV 3
  STATE a1, a2, a3: BOOLEAN
BEGIN
  { h-l > 2 | << out:= a1  AND  a2  AND  a3 >> ||
              atree(in, l, l+w-1, a1) ||
              atree(in, l+w, l+2*w-1, a2) ||
              atree(in, l+2*w, h, a3) } ||
  { h-l = 2 | << out:= in[l]  AND  in[l+1]  AND  in[l+2] >> }
END atree
```

Figure 2.14 Design description of an AND tree

Example: An Arbiter

An arbiter is a circuit that provides indivisible access to a shared resource, e.g., a bus or a peripheral. The arbiter described here is implemented as a binary tree in which all nodes (including the root and the leaves) are identical. The arbitration algorithm is based on passing a unique token around the tree. An external process using the arbiter is connected to a leaf of the tree, and it may use the resource only when that leaf has the token.

Each node of the tree has three pairs of connections (see figure 2.15), one for its parent and one for each of its children. A connection pair consists of two state variables, *req* and *gr*, standing for request and grant. Such a pair is used according to the following four-phase protocol.

1. A node requests the token by setting *req* to *TRUE*.
2. When *gr* becomes *TRUE*, the node has the token and may pass it down the tree.
3. The token is handed back by setting *req* to *FALSE*.
4. When *gr* becomes *FALSE*, a new request can be made.

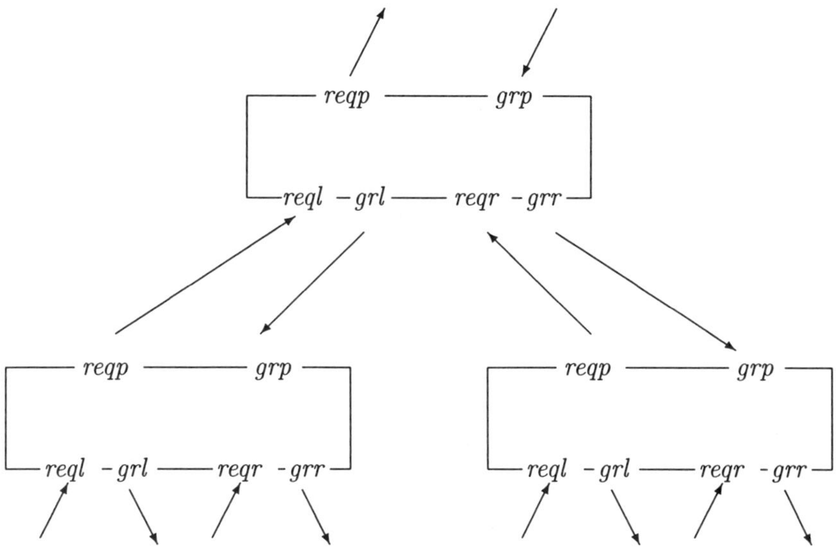

Figure 2.15 Two levels of arbiter tree

Figure 2.15 shows a few nodes and their interconnections. The state variables, *reqp* and *grp*, at a particular node are connected to *reql* and *grl*, or to *reqr* and *grr*, at the previous (higher) level of the tree (these connections are specified by the parameter passing). When both children of a particular node request the resource, it is given first to the left child; when that child releases the resource, it is given to the right child. A transition at the root copies *reqp* to *grp*, so that the root is able to propagate a grant immediately after it makes a request. The complete design description is shown in figure 2.16. This arbiter has no external connections (at the leaves), and it is quite simple to add these (see also exercise 2.8).

2.10 RESTRICTING STATE VARIABLES

Each named quantity has a scope which is the part of the design description where the definition of the name applies. The scope of a state variable is the cell in which it is declared, including named transitions, but excluding inner cell declarations. This means that variables declared in enclosing cells are not visible in a cell, unless explicitly passed as parameters. Functions and

```
CELL arbiter(grp, reqp: BOOLEAN; STATIC depth: INTEGER)
 STATE grl, reql, grr, reqr: BOOLEAN
BEGIN
 (* requestparent *)
  ≪ NOT grp AND (reql OR reqr) -> reqp:= TRUE ≫ ||
 (* grantleft *)
  ≪ grp AND reqp AND reql AND NOT grr -> grl:= TRUE ≫ ||
 (* grantright *)
  ≪grp AND reqp AND reqr AND NOTgrl AND NOTreql -> grr:= TRUE≫||
 (* doneleft *)
  ≪ NOT reql AND grl -> grl, grr, reqp:= FALSE, reqr, reqr≫||
 (* doneright *)
  ≪ NOT reqr AND grr -> grr, reqp:= FALSE, FALSE ≫ ||
 (* instantiate children *)
  {depth > 0 | arbiter(grl, reql, depth-1) || arbiter(grr, reqr, depth-1) }
END arbiter
```

Figure 2.16 Recursive description of arbiter

transitions are not visible in inner cells.

A state variable passed as an actual parameter can be read and written both by the instantiating and by the instantiated cell. This unrestricted use is the default, but it is also possible to restrict the use of state variables, for example, thus preventing them from being changed. There are three kinds of restrictions that can be specified on parameters and state variables

- External write: only external changes are allowed. Changes in the current scope or internal cells are not permitted. This restriction is only meaningful for formal parameters.
- Local write: only changes in the current scope are allowed. Changes in either internal or external cell instances are not allowed.
- Internal write: only changes in internal cell instances are allowed. Changes in the current scope or in external cells are not allowed.

A state variable or parameter which is visible (by the scope rules) can always be read. Syntactically, a restriction is specified as follows:

RESTRICTION
name: restriction

Here restriction is one of the reserved symbols *EXTERNAL*, *INTERNAL*, and *LOCAL*. The restrictions are listed among the declarations for a cell. For example, the arbiter cell might contain the following restrictions.

```
CELL arbiter(grp, reqp : BOOLEAN; STATIC depth : INTEGER)
  STATE grl, reql, grr, reqr : BOOLEAN
  RESTRICTION
    grp : EXTERNAL
    reqp, grl, grr : LOCAL
    reql, reqr : INTERNAL
```

These indicate that the formal parameter *grp* must not be written in the transitions of this arbiter cell. Similarly, *reql*, *reqr* must not be written in this cell; however, the internal cells to which they are passed as parameters may change their value. Finally, *reqp*, *grl*, and *grr* may be changed by the transitions of this arbiter cell, but not by external nor by internal cells. When a variable passed as an actual parameter is restricted to be externally or locally written, the corresponding formal parameter (in the cell instance to which the parameter is passed) must be declared to be external. The formal parameter *grp* obeys this rule.

2.11 OTHER CONSTRUCTS

This chapter has introduced a number of constructs for describing a transition system, in particular the cell concept and various forms of instantiations. These are the unique parts of Synchronized Transitions; however, a number of more conventional constructs are also convenient, for example, a type definition mechanism and functions. These concepts are not covered by this book. In the prototype tools supporting Synchronized Transitions there is a rudimentary type definition mechanism that is a subset of what is available in Pascal. This is explained in the documentation that is obtained with the tools supporting Synchronized Transitions.

The scope rules of Synchronized Transitions were mentioned in connection with cells where it is possible to hide state variables and transitions; however, the scope rules of statics, type names, etc. are not described in this chapter.

They are given in the language report contained in appendix A.

2.12 BACKGROUND

In [16] Dijkstra suggested a simple notation for describing a computation by guarded commands and nondeterministic selection. In this paper he pointed out that elimination of the ";" (sequencing) could simplify programs. This has directly and indirectly influenced many languages, for example, CSP [35]. In the author's Ph.D. thesis [58] a notation was used with many similarities to Synchronized Transitions, but a sequencing operator was used. In their book on UNITY [10] Chandy and Misra proposed a notation for describing concurrent computations based on guarded commands and with no sequencing operator. Like Synchronized Transitions there is operator for sequential composition in UNITY.

Models based on concurrency play a key role in many areas of computer science, including hardware design. Currently, the research emphasis seems to be on models based on synchronous message passing and first introduced by CSP [35]. UNITY and Synchronized Transitions differ significantly by using a model based on shared state variables. There is some similarity with the action systems proposed by Back and Kurki-Suonio [2]; however, their model has several concepts that are not found in Synchronized Transitions, for example, synchronization, explicit processes, and sequencing.

Synchronized Transitions differs from typical hardware description languages, HDLs, (e.g. VHDL [36] and ELLA [53]) in that preconditions (guards) determine the temporal behavior instead of event-list operations. This allows both synchronous and asynchronous circuits to be described using Synchronized Transitions, whereas other languages are only well-suited for synchronous circuits. Most HDLs express timing relationships implicitly (by counting clock cycles). In contrast, timing relationships in Synchronized Transitions are expressed explicitly in the preconditions of transitions (i.e., as a condition on the system state). Thus, high-level descriptions in Synchronized Transitions do not depend on details of the low-level implementation.

Constructs for modules or cells are found in most modern programming and hardware description languages [36]. The interface mechanisms in these languages differs; in some a parameter must be either input or output, e.g. Verilog, but not both as allowed in Synchronized Transitions. Recursive cell descriptions are also rare.

Synchronized Transitions has been developed to design hardware; however, there are also striking similarities with spreadsheets. The fundamental computing models are very similar namely, a collection of state variables (cells in spreadsheets) that are linked by transitions (formulas in spreadsheets) so updating one state variable may affect the values of other variables. This analogy was stressed in [9] and one motivation for initiating the work on UNITY. Compared to Synchronized Transitions, current spreadsheets have many limitations, for example, they are only two-dimensional, have very weak nesting mechanisms and addressing facilities. On the other hand, it might be these and other simplifications that explain their widespread use.

Hopefully, the examples of this chapter demonstrate that the computational model offered by Synchronized Transitions is both powerful and useful. Furthermore, it is quite different from the model offered by other programming and hardware description languages. However, the data model is very traditional and probably not adequate in the long run for describing large modular designs with many generic/parameterized parts. The constructs for specifying state variables and their types are similar to those found in Pascal. Since this language was introduced in 1971 many shortcomings have been discovered in particular with respect to parameterization and other abstraction mechanism. Synchronized Transitions share many of these shortcomings with Pascal. In the mean time more sophisticated concepts have been proposed for defining data types, for example, ML.

2.13 EXERCISES

Exercise 2.1 Let x, y, z, u, w be five state variables of type integer. Give a design describing a computation that orders the five state variables so that:

$$x \leq y \leq z \leq u \leq w$$

Exercise 2.2 Write a transition or cell for each of the following logic elements:

1. A two-input NAND-gate.
2. A transparent latch.
3. A flip-flop with preset and clear

Exercise 2.3 Write a design description that computes the greatest common divisor of three given numbers x, y, and z. What is the fixed point of the design?

Exercise 2.4 Let s : *ARRAY* $[1..n]$ *OF CHAR* be a string. Initially, the first m characters of s are in {'a'..'z'}, where $0 \leq m < n$. The last $n-m$ characters are the special end marker "." Write a design description that removes adjacent, duplicate characters, appending "." characters at the end as necessary. For example, "bbannnaanaa.." eventually becomes "banana.......".

Exercise 2.5 This problem deals with operations on a bit vector with n elements.

TYPE
  $range = [1..n]$
  $word$ = *ARRAY* $range$ *OF BOOLEAN*

Consider the following cell:

CELL $myst(w : word)$
BEGIN
  $\{\|i : [2..n] \mid \ll NOT\, w[i-1]\ AND\ w[i] \rightarrow w[i-1], w[i] := TRUE, FALSE \gg\}$
END

What is the fixed point for the computation described by cell $myst$?

Exercise 2.6 Consider a word consisting of n boolean values

TYPE
  $index = [0..n-1]$
  $word$ = *ARRAY* $index$ *OF BOOLEAN*

Write a cell, msb, in Synchronized Transitions. This cell should compute the highest index, r, where the value is true (most significant bit).

CELL $msb(b : word; r : index)$

It can be assumed that at least one of b's elements is true. What is the fixed point for the solution?

Exercise 2.7 Let W be a cost matrix of type $table$ as described in section 2.6. Describe a design that computes the number of stops in the cheapest route for getting between any pair of cities.

Exercise 2.8 Consider the arbiter discussed in section 2.9 and assume that the external connections are represented as two arrays:

TYPE $vector$ = *ARRAY* $[1..n]$ *OF BOOLEAN*
STATE $ereq, egr : vector$

Hence, $ereq[i]$, $egr[i]$ represent the request and grant signal to an external client i. Modify the recursive description of the arbiter presented in section 2.9 in such a way that the leaves of the arbiter tree are connected to the two vectors.

Exercise 2.9 Design an AND tree similar to the one presented in section 2.9 that also works if the number of leaves is not a power of 3.

Exercise 2.10 Design a cell $parity$ that computes the parity of a binary word.

TYPE
  $range = [1..n]$
  $word$ = *ARRAY* $range$ *OF BOOLEAN*

The result of a parity computation is a boolean that is true if there is an even number of bits of the word that are true and false otherwise.

t	x	y
1	0	0
1	1	0
0	1	0
0	0	0
1	0	0
1	0	1
0	0	1
0	0	0
1	0	0
1	1	0

Figure 2.17 Computation of a toggle

Exercise 2.11 Design a toggle element with an interface consisting of three variables t, x, and y. t alternates between true and false, and every time t becomes true, x or y should become true. Furthermore, x and y must alternate. t only becomes false after x or y has become true, when t is false both x and y must become false. The t can become true again, causing x or y to become true, etc. Figure 2.17 shows a representative sample of a computation.

3

FORMAL VERIFICATION

In the early phase of a design, many low-level details are left open while important high-level design decisions are made. Taking wrong or inconsistent decisions means extra work and increased design time. Since design time is often the most critical part of developing a new product, it is important to explore the properties of a design as early as possible. For example, verifying that different components of a design have a consistent view of their interface, or that a component has a particular property. Verification is the main topic of this chapter, but there are many different types of verification. One may, for example, want to show that a design,

- computes the correct function,
- operates at the desired speed,
- has an acceptable power consumption, or
- can be sold profitably.

Verifying these properties require different approaches, ranging from rigorous proofs to market surveys. The abstract description of a design has a role to play in many types of verification. However, in this chapter the emphasis is on design verification, where one shows a certain property of a design, for example, that communication follows a specified protocol, or that a given safety property holds. This kind of verification aims at the early phases of the design process. Another possibility is to do implementation verification where one shows that one design description is a refinement of another. This requires two design descriptions, for example, one describing a realization and one describing the abstract design. Implementation verification is, therefore, primarily aimed at a later phase of the design process and discussed further in chapter 6.

There are several ways of doing design verification, for example, by transforming the design description into a computer program that is executed on a number of test data sets (simulation). However, it is cumbersome and sometimes impossible to find test data covering all possible computations of a design, and if the test data does not exhaust all possible behaviors then errors in the uncovered parts are not revealed. In practice, one often has to settle for a nonexhaustive simulation. This is not the case for the formal verification technique described in this chapter. It corresponds to an exhaustive simulation without requiring a large set of test data and can, therefore, be a very effective way of verifying a design description.

A high-level description of a design is an abstraction of its physical behavior. It is important to realize that formal verification deals with the abstraction and *not* with the physical realization. A formal verification of a requirement, like the safety of a traffic light, is not an absolute guarantee against malfunctioning, for example, failure caused by a car hitting the traffic light controller or a power failure. If either of these happens, the formal description is no longer a model of the physical realization, and hence, properties verified from the description may no longer hold. The same applies to simulation which is also based on a model of the physical reality. If the model is not an adequate abstraction, a simulation, no matter how exhaustive, does not provide any guarantee against malfunctioning. *Verification ensures consistency between two descriptions*, for example, one modeling the behavior of a design and another modeling a requirement. It is outside the realm of formal methods to ensure that a model adequately reflects physical phenomena such as light or changing voltages. Despite these reservations, formal verification can be a very powerful tool, but it is important to realize that it can never become an absolute guarantee.

The first part of this chapter shows how to formalize design properties as assertions (logical expressions). It is then shown how such formally stated properties are verified. The verification technique is first explained for designs consisting of a single cell. Furthermore, it is shown how to use a mechanical theorem prover as a design automation tool aimed at the high-level design. The last part of the chapter shows how to verify modular designs with nested cells. This is done by verifying the cells separately, which means that verification of a large design is broken into many independent parts.

3.1 INVARIANTS AND PROTOCOLS

This section explains how properties of a design are expressed as invariants and protocols which are assertions (logic expression) that express constraints on the transition system. In section 3.2, it is shown how to verify that a design description is consistent with such assertions. Constraints on the states are called invariants, and constraints on the transitions are called protocols.

- An *invariant* defines a stable subset of the state space containing the initial state. A subset is stable, if there are no transitions from a state within the subset to one outside. An invariant is written as a boolean expression (that is true for all members of the subset and false otherwise). For the traffic light controller (introduced in section 1.3) the invariant:

  $$EWlight = red \quad OR \quad NSlight = red$$

  expresses the requirement: *the light is always red in at least one of the two directions.* This assertion, which is an example of an invariant, must be satisfied in all states of the computation.

- A *protocol* defines a set of pairs of states that must contain all state pairs, $(pre, post)$, specified by the transitions. A protocol is written as a predicate using names of state variables extended with the suffixes $.pre$ and $.post$. So, $v.pre$ refers to the value of the state variable before the state change (the pre-state), and $v.post$ refers to the value of the state variable immediately after (the post-state). In the traffic light controller, signals never change from $green$ to red. This constraint is expressed by the following protocol:

  $$NOT(EWlight.pre = green \quad AND \quad EWlight.post = red)$$

A design description may contain invariant and protocol declarations, which are indicated with the reserved symbols $INVARIANT$ and $PROTOCOL$ and written among the declarations, for example:

STATE $EWlight, NSlight$: *BOOLEAN*

INVARIANT $EWlight = red \ OR \ NSlight = red$

PROTOCOL $NOT(EWlight.pre = green \ AND \ EWlight.post = red)$

Invariant and protocol declarations have no influence on the dynamic behavior of a design. If they are removed from a description, it specifies *exactly* the same computation. So the protocols and invariants provide redundant information. This is quite similar to the use of types in high-level languages, for example Pascal, where the type definition of a variable in a syntactically correct program is redundant. However, this redundancy is used to check consistency between the declaration and use of a variable. Similarly, protocols and invariants are used to check the consistency of the transitions. Types are usually checked syntactically by the compiler; this is not possible for invariants and protocols which require a more complicated check. There are several possibilities, for example:

- by a simulation, where the design is executed on specific input data,
- by an informal, but convincing argument,
- by enumerating and checking all possible states and transitions,
- by a formal proof.

The emphasis in this chapter is on verification based on formal proofs, either done by hand or supported by a mechanical theorem prover. Before going into the details of the verification technique, a number of examples of invariants and protocols are given.

Example: Greatest Common Divisor

Section 2.5 shows a design for computing the greatest common divisor of two natural numbers X and Y using Euclid's algorithm. Two state variables x, y, initialized with the values of X, Y, are gradually reduced until they contain the value of the greatest common divisor. The invariant of this algorithm is:

$$\mathit{INVARIANT}\quad gcd(x, y) = gcd(X, Y)$$

Example: No Adjacent Duplicates

Let s be a string of characters (represented in some numerical representation), that is, a state variable of type: $ARRAY\ [1..n]\ OF\ char$. Initially, the first m characters of s, where $0 \leq m \leq n$, are in {'a'..'z'}, and the last $n-m$ characters are the special end marker: EOS. The problem is to remove adjacent duplicate characters from s, appending EOS characters at the end as necessary. For example, "bbannnaanaa" eventually becomes "banana".

Consider three adjacent elements, $s[i], s[i+1]$, and $s[i+2]$. If $s[i]$ and $s[i+1]$ contain the same character, the one in $s[i+1]$ is removed; otherwise $s[i+1]$ is left unchanged.

$$\ll s[i] = s[i+1] \mathrel{-\!\!>} s[i+1] := s[i+2] \gg$$

```
STATIC EOS = ...
TYPE
    char = INTEGER (* ASCII representation of characters *)
    range = [1..n]
    string = ARRAY range OF char
CELL nodup(s: string)
    TRANSITION copy(a, b, c: char)
        ≪ a = b -> b:= c ≫
BEGIN
    {|| i: [1..n-2] | copy(s[i], s[i+1], s[i+2]) }||
    ≪ s[n-1] = s[n] -> s[n]:= EOS ≫
END nodup
```

Figure 3.1 Removing adjacent duplicates

A similar transition operates at every position of the string; this is specified by a quantified instantiation, describing $n-2$ transitions, each responsible for removing possible duplicates in one position of the string (concurrently).

$$\{\ i : [1..n-2] \mid \ll s[i] = s[i+1] \text{ -> } s[i+1] := s[i+2] \gg \}$$

The transition described below places an end marker, EOS, at the end of the string.

$$\ll s[n-1] = s[n] \text{ -> } s[n] := EOS \gg$$

A complete design is shown in figure 3.1. An example of an invariant for this design is:

$$\textit{INVARIANT FORALL } h, j : range.\ (s[h] = EOS\ AND\ j > h) \Rightarrow s[j] = EOS$$

This states that *EOS* is only found in the rightmost part of the string, i.e., no end markers appear in between letters. The invariant is rather weak since it does not constrain the elements in the leftmost part of the string before the end marker. It is certainly possible to formulate stronger invariants, for example, one which states that the number of different characters before the end markers remains constant (see exercise 3.4). However, there is no obligation for the designer to formulate the strongest possible invariants. Sometimes, a weak invariant gives a sufficient assurance; in other cases, a strong invariant is the best choice. It is up to the designer to formulate invariants that are strong enough to ensure against the errors he is concerned about.

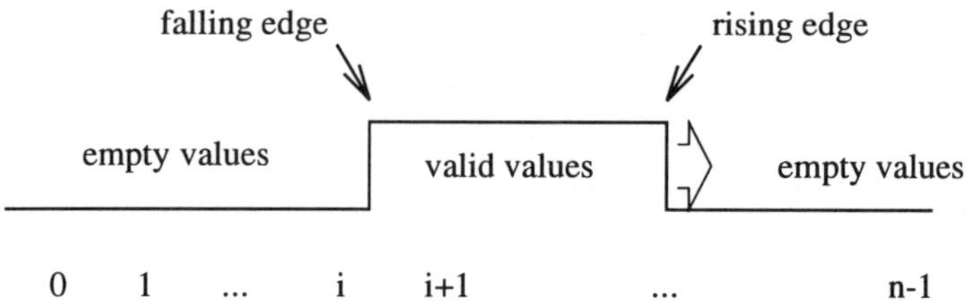

Figure 3.2 Wave of computations

Example: Maximum

Let a be a vector of integers for which the maximal value should be computed.

TYPE
  $range = [0..n-1]$
  $vector$ = *ARRAY* $range$ *OF INTEGER*
STATE
  a : $vector$
  m: *INTEGER*

The objective is to describe a computation that achieves, $m = max_{0 \leq j < n}\ a[j]$. As it is often the case, an invariant is obtained by weakening the objective, e.g., $m \leq max_{0 \leq j < n}\ a[j]$. However, to be a maximum, m must also be one of the elements of the array: $\exists i : a[i] = m$. These two assertions are combined to form the following invariant:

INVARIANT
  $\{\ OR\ j : [0..n-1] \mid m \leq a[j]\}\ \ AND\ \ \{\ OR\ j : [0..n-1] \mid a[j] = m\}$

Example: The Iterative Ring

In the iterative ring, introduced in section 2.8.3, computation progresses as a wave traveling around the ring. A series of consecutive valid values constitutes the wave that represents one datum in the ring. The rising edge of the wave (valid values) corresponds to computing a new result; the falling edge (empty value) corresponds to emptying in preparation for the next computation. This is illustrated in figure 3.2. The falling edge is described with the predicate, *drop*:

FUNCTION $drop$(*STATIC* i: *INTEGER*): *BOOLEAN*
  RETURN $(v[i] = E)\ AND\ (v[(i+1)\ MOD\ n] \neq E)$

A proper computation must ensure that *the rising edge of the wave never overtakes the falling edge* (i.e., attempting to perform a computation on a stage that is not emptied from the previous computation), and *that the falling edge never overtakes the rising edge* (i.e., attempting to empty a stage before the result is received by the succeeding stage). The following invariant requires that there is exactly one wave.

INVARIANT
FORALL $j, k : index \; . \; drop(j)$ *AND* $drop(k) \Rightarrow j = k$
EXISTS $k : index \; . \; drop(k)$

The first part ensures that there is at most one falling edge and the second part that there is at least one.

Example: The Arbiter

The arbiter described in section 2.9 provides indivisible access to a shared resource. A number of external processes, using the resource, are connected to the arbiter with a pair of state variables, req and gr (request and grant). There is a separate pair, req, gr, for each external process and for each node of the arbiter tree (see figure 2.15). The arbiter must never grant access to two nodes simultaneously, and this is expressed with the invariant:

INVARIANT NOT $(grr$ AND $grl)$

To operate correctly, the arbiter may assume that all request/grant pairs follow the four-phase protocol described in section 2.9.

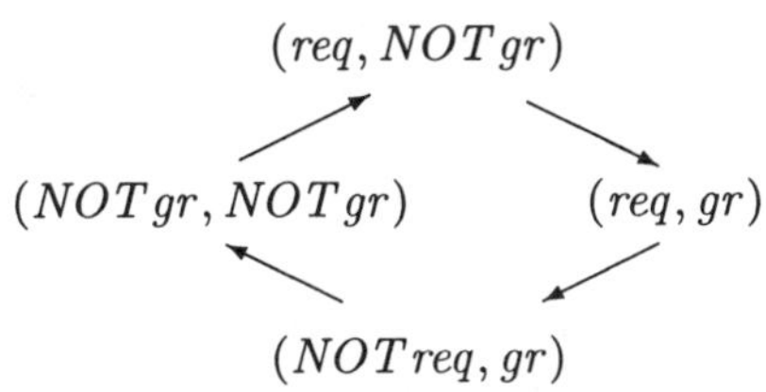

This is expressed formally with a protocol as follows:

PROTOCOL
$req.post \neq gr.post \Rightarrow req.post \neq req.pre$
$req.post = gr.post \Rightarrow gr.post \neq gr.pre$

3.1.1 Expressing a Transition as a Predicate

To do formal verification, it is useful to have a transition expressed as a predicate in a notation similar to the one used for invariants and protocols. A transition, t, relates a pre and a post-state, which can be expressed by a predicate, $t(pre, post)$, which is true if and only if the transition leads to state $post$ when started in state pre. Consider, for example the transition

$\ll$ *(EWlight = green) AND CarOnNS –> EWlight:= yellow* $\gg$

This can also be expressed as the predicate

(EWlight.pre = green) AND CarOnNS.pre AND
(EWlight.post = yellow) AND
(CarOnNS.pre = CarOnNS.post) AND (CarOnEW.pre = CarOnEW.post) AND
(NSlight.post = NSlight.pre)

The first line corresponds to the precondition, the second to the assignment, and the two last lines are necessary in order to express that only the state variable *EWlight* may change its value. This formalization of a transition as a predicate is used several times in the rest of the book, for example in the following section.

3.2 VERIFICATION OF INVARIANTS AND PROTOCOLS

This section presents a formal technique for verifying invariants and protocols. To introduce the technique, it is first explained for designs consisting of a single cell. In section 3.4 it is shown how to extend the verification technique to modular designs with many cells.

The verification must establish that all states of a computation satisfy the invariant, and that all transitions meet the protocols. Doing this explicitly by enumerating and checking all possible states and transitions could become very laborious. Instead, an inductive technique is used where the complexity of the verification is linear in the number of transition descriptions and independent of the size of the state space or length of the computation. This approach was originally proposed by Floyd for software verification [22].

Assume that I is an invariant and P a protocol of the design D that is going to be verified.

- The *base* of the induction must establish that the initial state, $init$, satisfies the invariant, i.e., showing that $I(init)$ is satisfied.
- The *induction step* establishes that the post-state of each transition satisfies the invariant and protocol, assuming that the pre-state satisfies the invariant. It must be shown that all transitions, $t \in D$, satisfy the implication:

  C: $I(pre) \wedge t(pre, post) \Rightarrow I(post) \wedge P(pre, post)$

  This is called the correctness condition (**C**).

The correctness condition suggests that the verification is done by *considering the transitions one at a time*, and showing the implications separately for each transition. Note that the verification of a design is thereby reduced to a number of independent steps, one for each transition. This is important both for manual and mechanical verification, because a large design is handled by a number of (relatively) simple proofs instead of a few large and complicated ones. A transition is said to maintain an invariant if the implication in **C** can be shown. Usually, it is not explicitly mentioned that formally the verification technique is induction. The correctness condition, **C**, is refined later in this chapter to enable a modular verification of designs with a hierarchy of cells.

Example: Verifying the Traffic Light Controller

Figure 3.3 shows the traffic light design including the invariant discussed in section 3.1. The initial state satisfies the invariant, because $NSlight$ is set to red. Furthermore, it must be shown that the invariant is maintained by each of the four transitions. The first is:

$$\ll EWlight = green \ AND \ CarOnNS \ {-}{>} \ EWlight := yellow \gg$$

To show that this transition maintains the invariant, assume that the invariant is satisfied before an execution of the transition. The precondition allows us to conclude that $EWlight = green$, hence $NSlight = red$ (otherwise the invariant is not satisfied in the pre-state). But $NSlight$ is not changed by the transition, so $NSlight = red$ must also hold in the post-state. Therefore, the first transition maintains the invariant. The second transition is:

$$\ll EWlight = yellow \ {-}{>} \ EWlight, NSlight := red, green \gg$$

It cannot violate the invariant, because $EWlight$ is explicitly given the value red, and therefore, the invariant is satisfied after execution of the transition. The last two transitions are symmetrical, and therefore the arguments that show that they maintain the invariant are similar. The invariant has now been verified, since it is satisfied by the initialization and maintained by all four transitions.

```
CELL trafficlight
  STATIC  green = 0  yellow = 1  red = 2
  STATE
    EWlight, NSlight: INTEGER
    CarOnEW, CarOnNS: BOOLEAN
  INVARIANT  EWlight=red  OR  NSlight=red
  INITIALLY  EWlight = green   NSlight = red
BEGIN
  ≪ EWlight = green  AND  CarOnNS  ->  EWlight:= yellow ≫ ||
  ≪ EWlight = yellow  ->  EWlight, NSlight:= red, green ≫     ||
  ≪ NSlight = green  AND  CarOnEW  ->  NSlight:= yellow ≫  ||
  ≪ NSlight = yellow  ->  NSlight, EWlight:= red, green≫
END trafficlight
```

Figure 3.3 Traffic light controller

Example: Verifying the Greatest Common Divisor

Figure 3.4 shows the complete design description including an invariant. The initial values of the state variables x and y are X and Y respectively. Therefore, the initial state satisfies the invariant. Replacing x by $x - y$ when $x > y$ does not change $gcd(x, y)$, hence the first transition maintains the invariant, and the argument for the second one is symmetrical. It has therefore been verified that $gcd(x, y) = gcd(X, Y)$ is an invariant.

```
STATE  x,y: INTEGER
INVARIANT  gcd(x, y) = gcd(X, Y)
INITIALLY  x=X   y=Y
BEGIN
   ≪ x > y  ->  x:= x-y ≫   ||   ≪ x < y  ->  y:= y-x ≫
END
```

Figure 3.4 Greatest common divisor of X and Y

3.3 MECHANICAL VERIFICATION

There are many CAD tools available to help designers master the complexity of large designs. Most of these tools are aimed at a low level, e.g., at the layout or transistor level. There is a similar need for tools to assist in doing high-level design. This section describes such tools that are used for verification. To verify a large design, it is necessary to check many instances of the correctness condition, **C** (see page 55). It becomes quite tedious to do this by hand as it was done in section 3.2.Such checks rarely require much insight or mathematical sophistication. Furthermore, the checking must be redone every time a design description is changed. By using a mechanical theorem prover, it is possible to reduce the routine work needed to do formal verification.

3.3.1 Mechanical Theorem Proving

There is a number of mechanical theorem provers available. They differ significantly, for example, in the expressive power of the notation allowed for stating a conjecture. However, the notion of proof used in all the currently available mechanical theorem provers is rather similar. A proof is a series of reductions transforming the conjecture into a formula known to be true (or false). As an example, consider showing the simple conjecture:

$$(2 * 2) = (2+2).$$

Shown below is a series of reductions done by a mechanical theorem prover. In this reduction $s(0)$ denotes 1, and $s(s(0))$ denotes 2, etc.

1. $(2 * 2) = (2+2)$
2. $(s(s(0)) * 2) = (2+2)$
3. $(s(s(0)) * s(s(0))) = (2+2)$
4. $((s(s(0)) * s(0))+s(s(0))) = (2+2)$
5. $(2+2) = s((s(s(0)) * s(0))+s(0))$
6. $(s(s(0))+2) = s((s(s(0)) * s(0))+s(0))$
7. $(s(s(0))+s(s(0))) = s((s(s(0)) * s(0))+s(0))$
8. $s((s(s(0)) * s(0))+s(0)) = s(s(s(0))+s(0))$
9. $s(s(s(0))+s(0)) = s(s((s(s(0)) * s(0))+0))$
10. $s(s((s(s(0)) * s(0))+0)) = s(s(s(s(0))+0))$
11. $s(s(s(s(0)) * s(0))) = s(s(s(s(0))+0))$
12. $s(s((s(s(0)) * 0)+s(s(0)))) = s(s(s(s(0))+0))$
13. $s(s(s(s(0))+0)) = s(s(s((s(s(0)) * 0)+s(0))))$
14. $s(s(s((s(s(0)) * 0)+s(0)))) = s(s(s(s(0))))$

15. $s(s(s(s((s(s(0)) * 0)+0)))) = s(s(s(s(0))))$
16. $s(s(s(s(s(s(0)) * 0)))) = s(s(s(s(0))))$
17. $s(s(s(s(0)))) = s(s(s(s(0))))$
18. *true*

Each line represents a small reduction of the previous line using simple rules like $x * y = x * (y-1)+x$. A series of such reductions yield the equation $s(s(s(s(0)))) = s(s(s(s(0))))$ which is known by the theorem prover to be true.

The reductions can either be built into the theorem prover or supplied by the user. For example, once a conjecture has been verified, it can be used to reduce other conjectures. This mechanical notion of a proof as something consisting of a finite number of applications of reductions is somewhat more restricted than the notion of proof usually found in mathematics. However, verification of most functional properties of designs does not require deep mathematical insight and, therefore, the mechanical notion of proof is usually adequate.

The theorem prover LP [25] has been used in experiments with mechanical verification of designs in Synchronized Transitions, and section 3.3.3 sketches how this is done. LP is a theorem prover for a subset of multi-sorted first-order logic. It is designed to work efficiently on large problems and to be used by relatively naive users. Among its uses are debugging formal specifications written in Larch [27], reasoning about algorithms which involve concurrency [28], and establishing the correctness of hardware designs which is the topic of this chapter.

LP is intended primarily for use as an interactive proof assistant, not as a fully automatic theorem prover. *Its design is based on the assumption that initial attempts to state conjectures correctly, and then to prove them, usually fail.* As a result, LP is designed to carry out routine (and possibly lengthy) steps in a proof automatically and to provide useful information about why proofs fail, if and when they do. Unlike the Boyer-Moore prover [4, 5], it does not use heuristics to formulate additional conjectures in a search for a proof that may not exist. Strategic decisions, such as trying induction, must appear as explicit LP commands (either entered by the user or generated by an application-specific front-end such as the translator ST2LP discussed in section 3.3.4). When proof attempts fail, LP provides quick feedback; in fact, if and when time-consuming proof attempts occur, they are generally a sign that something is wrong.

3.3.2 LP's Logical System

Operationally, the basis for proofs in LP is a logical system that contains a small set of built-in axioms concerning booleans and conditionals, together with user-supplied axioms and theorems which include the following kinds:

- *Equations* of the form $t_1 == t_2$, where t_1 and t_2 are terms.
- *Rewrite rules* of the form t_1 -> t_2, which have the same logical meaning as equations, but which behave differently operationally. LP converts equations into rewrite rules by various ordering procedures. It uses rewrite rules to reduce conjectures and facts in its logical system to normal forms. Equality is decided by comparing such normal forms.
- *Deduction rules*, which LP uses to infer new equations from existing equations and rewrite rules. For example, the built-in deduction rule:

  when $not(x) == true$ *yield* $x == false$

  enables LP automatically to replace the rewrite rule:

  not(red = green) -> true

  by the simpler rule:

  red = green -> false

Details concerning LP can be found in [26, 33].

3.3.3 Using LP for Verification

This section illustrates how LARCH PROVER can be used to verify a simple invariant. Usually it is not necessary to have a detailed knowledge of LARCH PROVER in order to verify simple designs using the tools mentioned in section 3.3.4. It is sketched how LP can be used for verifying the invariant of the traffic light controller.

The goal of an invariance proof is to show a number of implications as stated in the correctness condition given in section 3.2.

C: $I(pre) \wedge t(pre, post) \Rightarrow I(post) \wedge P(pre, post)$

To use LP for showing such implications, it is necessary to formalize, the notion of state, invariants, protocols, and transitions.

States

First, all nonboolean values that state variables can take are formalized by introducing appropriate sorts (LP's term corresponding to a type). For example, by using *declare* statements

$$declare\ sort\ Color$$
$$declare\ operators\ green, yellow, red : -> Color$$

Colors are introduced as constants. LP assert statements are used to express properties (axioms) of these constants.

$$\begin{array}{l} assert \\ \quad not(green = red) \\ \quad not(yellow = red) \\ \quad not(green = yellow) \end{array}$$

State variables are formalized as mappings (operators) of appropriate sorts:

$$declare\ sorts\ ColorVar, BoolVar$$
$$declare\ operators\ EWlight, NSlight : -> ColorVar$$
$$declare\ operators\ CarOnEW, CarOnNS : -> BoolVar$$

The pre- and post-states are two particular values of sort $State$. The dot operator, e.g., $CarOnEW.pre$, is formalized as a mapping with two arguments, a state variable and a state.

$$\begin{array}{llll} declare\ sort\ State & & & \\ declare\ operators & & & \\ \quad pre, post: & & -> & State \\ \quad .\!: & ColorVar, State & -> & Color \\ \quad .\!: & BoolVar, State & -> & Bool \end{array}$$

Invariants

Invariants are formalized as unary predicates, defined by a single equation:

$$declare\ operator\ I : State -> Bool$$
$$declare\ variable\ s : State$$
$$assert\ I(s) == EWlight.s = red \mid NSlight.s = red$$

The operator | is boolean disjunction, *OR* (& is boolean conjunction, *AND*).

Transitions

Transitions are formalized as boolean relations on the pre and post-states (see section 3.1.1). It turns out that LP can better utilize assumptions if they are expressed as deduction rules, for example:

```
declare operators Action, Action1, Action2, Action3, Action4: -> Bool
assert
  when Action yield
    Action1 | Action2 | Action3 | Action4
  when Action1 yield
    EWlight.pre = green, CarOnNS.pre          % Precondition
    EWlight.post = yellow                     % Action
    U(NSlight), U(CarOnNS), U(CarOnEW)        % Unchanged
  when Action2 yield
    ⋮
```

The first deduction rule asserts that there are four transitions, any one of which can be selected (nondeterministically) for execution. The second deduction rule causes LP to deduce six conclusions describing the transition whenever it discovers that *Action1* is true, for example, whenever the user asserts *Action1*. The first two conclusions assert that the precondition: $EWlight.pre = green \wedge CarOnNS.pre$ is true in the pre-state, the third that: *EWlight* has a new value in the post-state, and the last three that the values of the other state variables do not change. The properties of the (overloaded) operator *U* are axiomatized by the declarations and equations

```
declare operators U : ColorVar -> Bool, U : BoolVar -> Bool
declare variables c : ColorVar, b : BoolVar
assert
  U(c) == c.post = c.pre
  U(b) == b.post = b.pre
```

Verification Condition

To verify that the invariant is maintained, a conjecture is introduced and LP is directed to prove it by assuming the hypothesis of the implication and then doing a case analysis:

```
prove (I(pre)&Action) ⇒ I(post)
    resume by ⇒
    resume by cases Action1, Action2, Action3, Action4
```

LP carries out the details of the proof without any further user interaction. In response to ***resume by*** $\Rightarrow$, LP tries to prove the subgoal $I(post)$ using the additional axiom: $I(pre) \& Action$, from which it deduces both $I(pre)$ and $Action1 \mid \cdots \mid Action4$. In response to: ***resume by cases ...***, LP first verifies that the cases are mutually exhaustive; then it proceeds to prove the subgoal $I(post)$ from each of the additional axioms $Action1, \ldots, Action4$. The first case is typical. LP applies the deduction rule defining *Action1* to derive six new equations, which it orients into rewrite rules. The first of these, together with the rewrite rule: *green* = *red* –> *false* deduced from the axiomatized properties of colors, reduce $I(pre)$ to $NSlight.pre = red$; LP transforms this equation and $U(NSlight)$ into rewrite rules that reduce the subgoal $I(post)$ to an identity.

Proofs such as this illustrate LP's rewriting strategy. By reducing axioms as well as conjectures to normal forms, LP is often able to deduce enough additional rewrite rules to reduce a correct conjecture to an identity.

3.3.4 Automatic Translation

To make use of LP, the verification conditions must be formulated in a way similar to the one shown in section 3.3.3. Doing this by hand is as tedious as carrying out the proof. Therefore a translator has been constructed that automatically transforms a design description with invariants and protocols to verification conditions for LP. The translator is called ST2LP; it extracts appropriate declarations for state variables, axioms defining transitions, protocols, invariants, and verification conditions from design descriptions. The translator takes care of routine details, such as generating clauses for unchanged state variables, and it generates LP commands similar to the ones shown in section 3.3.3. Hence, the designer need not know all the details about LP to verify designs in Synchronized Transitions. The translator also generates fruitful idioms for the proof; when formalizing invariance proofs, the translator generates an LP command to initiate a proof by cases on the transition selected for execution. As a result of these idioms, LP often verifies invariants of designs directly from the output of the translator without any manual assistance. This is for example the case for the traffic light controller: if the design description shown in figure 3.3 is given to ST2LP, it produces a list of verification conditions that LP can prove without any manual intervention.

In section 3.4 it is explained how to verify modular designs with many cells utilizing ST2LP and LP. This approach has been used to verify a number of nontrivial designs in addition to all the small examples described in this chapter.

The tools provide a significant assistance in particular with the routine details; however, formal verification of nontrivial designs still requires a substantial effort. For designs going through many iterations most effort is needed when a property is verified for the first time. Mechanized proofs are easily repeatable. This often means that re-checking a property after a small change is easy. Only further use will show whether designers find this effort worthwhile. However, it is the author's experience that whenever complexity shows up, formal analysis and verification are valuable.

The tools LP and ST2LP are used for a number of other verification purposes:

- checking fixed points (see section 3.3.5),
- verifying refinements (implementation verification) (see chapter 6), and
- checking restrictions imposed by the chosen realization technique (see section 5.4.1 and 7.3.1).

These are all based on the same concepts as the design verification discussed in this chapter. It stresses the point made in the introduction: once a formal description of a design has been made, it can be scrutinized from many different angles providing a firm foundation for further use of the design description.

Example: A Modified Traffic Light Controller

Consider once more the traffic light controller. Figure 3.5 shows a slightly modified version of the design shown in figure 3.3. The only difference between the two designs is that an extra state variable, d, has been introduced to represent the direction in which cars are currently allowed to pass. The same invariant $EWlight = red\ OR\ NSlight = red$ is satisfied by all the states encountered by the modified design. It is, however, not possible to verify the invariant using the technique described above. This is because, the invariant does not carry any information about d, hence, it is no help to assume that the invariant holds in the pre-state. To verify the invariant, it is necessary to strengthen it, i.e., to make it say more, for example:

$$INVARIANT\ (NOT\, d \Rightarrow EWlight = red)\ AND\ (d \Rightarrow NSlight = red)$$

It is simple to verify this invariant using ST2LP and LP, and since

$$\begin{aligned}(NOT\, d \Rightarrow EWlight = red)\ AND\ (d \Rightarrow NSlight = red)\\ \Rightarrow (EWlight = red\ OR\ NSlight = red)\end{aligned}$$

```
CELL trafficlight
  STATIC  green = 0  yellow = 1  red = 2
  STATE
    EWlight, NSlight: INTEGER
    CarOnEW, CarOnNS: BOOLEAN
    d: BOOLEAN
  INVARIANT  EWlight=red   OR   NSlight=red
  INITIALLY  EWlight = green   NSlight = red
BEGIN
   ≪ EWlight = red  AND  d  –>  EWlight:= green ≫  ||
   ≪ EWlight = green  AND  CarOnNS   –>  EWlight:= yellow ≫  ||
   ≪ EWlight = yellow  –>  EWlight, d:= red, FALSE  ≫  ||
   ≪ NSlight = red  AND  NOT d  –>  NSlight:= green ≫  ||
   ≪ NSlight = green  AND  CarOnEW  –>  NSlight:= yellow ≫  ||
   ≪ NSlight = yellow  –>  NSlight, d:= red, TRUE ≫
END trafficlight
```

Figure 3.5 Traffic light controller

the assertion $EWlight = red\ OR\ NSlight = red$ is satisfied for all states of the computation.

This simple example illustrates a quite common phenomenon; to verify an invariant, it is necessary to make it stronger. At first, it might be a little surprising that strengthening an invariant makes it easier to verify, since fewer states meet the stronger invariant. The reason is that the verification technique allows one to *assume* that the invariant is satisfied in the pre-state, hence, a stronger invariant also means a stronger assumption. There is, however, a limit where further strengthening does not help, because the resulting assertion no longer is an invariant. In practice, finding the right balance between assuming enough, but not too much can require some experimentation. The theorem prover can provide significant help because it can point out a counterexample to the proposed invariant. Such a counterexample is often a help in strengthening the invariant. The example of a switch presented in section 7.3.6 illustrates this.

3.3.5 Checking Fixed Points

The fixed point of a design can be stated explicitly in the design, by giving an assertion that must hold at a fixed point, for example:

FIXPOINT $x = y$

Such a fixed point assertion plays a similar role to an invariant assertion, it provides redundant information that does not affect the computation. Through formal verification it can be shown that the redundant information is consistent with the rest of the design.

Figure 3.6 shows a design description of the greatest common divisor computation including a fixed point specification. Using ST2LP and LP it can be shown that the fixed point specification is consistent with the transitions. Further details are given in [51].

3.4 VERIFICATION OF MODULAR DESIGNS

The examples given above were all small designs consisting of a single cell. This section describes how to verify a modular design with a hierarchy of cells. For such designs, it is important to enable the designer to do the verification in

```
STATE x,y: INTEGER
INVARIANT gcd(x, y) = gcd(X, Y)
INITIALLY x=X  y=Y
FIXPOINT x=y
BEGIN
  ≪ x > y -> x:= x-y ≫ || ≪ x < y -> y:= y-x ≫
END
```

Figure 3.6 Greatest common divisor with fixed point specification

small pieces, for example, verifying the cells one at a time separately from the verification of the composition. Furthermore, it should be sufficient to verify a cell once, rather than doing a separate proof for each instantiation.

A modular design describes a transition system where the transitions are divided into disjoint subsets, one subset for each cell instance. The state changes specified by a particular transition might violate invariants and protocols of the cell instance to which the transition belongs. However, protocols or invariants of other instances might also be violated by changing the value of a state variable passed as a parameter. In principle, it is therefore necessary to verify all transitions globally, i.e., to show that every transition maintains the invariants and protocols of *all* cell instances. This is not practical for large designs, because the verification effort becomes prohibitive. Below, a localized verification technique is introduced which requires less effort. It is introduced in two steps to simplify the explanation.

Instance verification. In the first step, localized verification conditions are formulated using the instance tree introduced in section 2.8.2. This leads to a modular verification where cells are verified separately which means that they can be verified one at a time. In many cases, this results in a number of similar verification conditions; for example, all the conditions generated from a quantified instantiation are almost the same. In the second step, it is explained how these similar conditions are merged.

Syntax directed verification. In the second step, the localized verification conditions are derived directly from the syntactic design description instead of the instance tree. This makes it possible to merge similar conditions, for example, all the conditions generated from a quantified instantiation are merged.

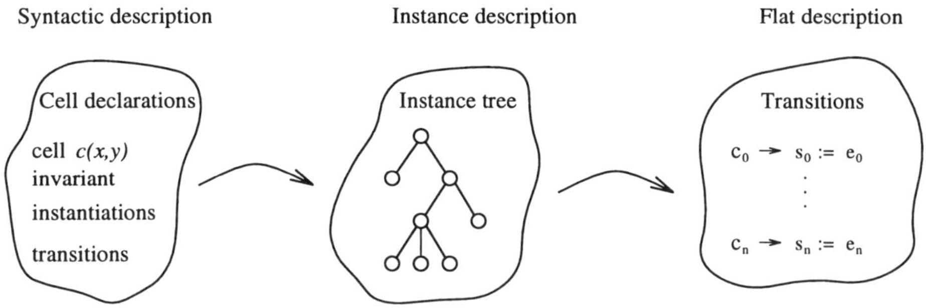

Figure 3.7 Three description levels

Figure 3.7 illustrates the three different views of a design description.

The iterative ring introduced in section 2.8.3 can be used to illustrate the different approaches. To verify the invariant presented in section 3.1, it must be shown that it is maintained by every transition in each instance of the cell *stage*. The instance tree for the design contains one instance of the cell *ring*, which is the root node. This instance has n subcell instances of the cell *stage*. The instance verification consists of verifying each of the $n+1$ cell instances separately. This separation is already a significant simplification compared to a global verification. However, the $n+1$ cell instances are very similar, and they can all be verified by a single proof using the syntax directed verification.

The benefit of the localized technique is illustrated by comparing the number of verification conditions needed for the different approaches.

- The global correctness condition is based on the flat description (see figure 3.8) where all cell instantiations and quantification have been expanded. This condition requires that every transition in every cell instance is shown to maintain the protocols and invariants in *all* other cell instances, including those operating on completely disjoint state variables. The number of verification conditions is proportional to the total number of instantiated transitions, and each condition involves all invariants and protocols of the entire design description.
- If the verification conditions are generated from the instance tree, the number of verification conditions is proportional to the size of the instance tree,

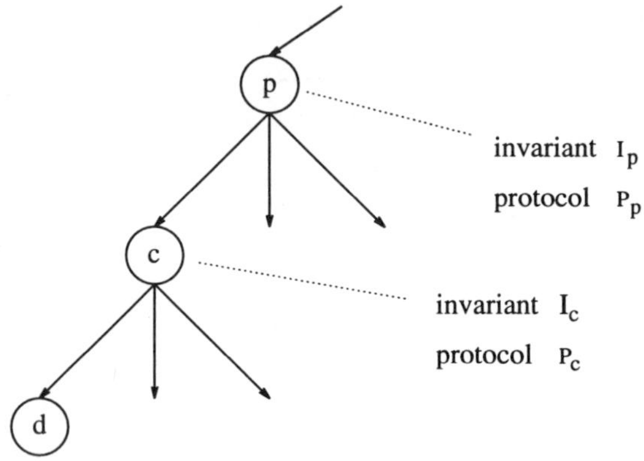

Figure 3.8 Instance tree

i.e., the number of cell instances. However, each verification condition only involves a few protocols and invariants of neighboring cell instances.

- The instance tree contains a number of similar cell instances (e.g., the instances of the cell *stage*). The verification conditions are very similar and can be merged to a few generic conditions. In this way, the number of verification conditions is reduced from a number proportional to the design parameters, to a number only proportional to the size of the syntactic design description.

In this comparison, efficiency is measured as the number of verification conditions, without regard to the size/complexity of each of these. As illustrated by the examples given later in this chapter, the size of the verification does not explode. Hence, the verification effort is reduced by using the syntax directed conditions.

3.4.1 Instance Verification

Figure 3.8 shows a generic instance tree that is used to introduce the localized verification technique. It has a parent cell instance, p, and a child cell instance, c. Each cell instance has an invariant and a protocol. In a particular case, one or both of these might not be present, because the designer has chosen not to formulate an invariant or a protocol. Formally, omitting a protocol or an invariant is the same as using the nonconstraining predicate $TRUE$. Each cell instance also has a set of transitions which are not shown in the figure.

The following abbreviations are used to describe the verification conditions for the localized verification technique.

$$\begin{aligned}
Assumption_a &\equiv I_a(pre) \\
Assumption_{a,b} &\equiv I_a(pre) \wedge I_b(pre) \\
Transition_a &\equiv \exists\, t \in T_a : t(pre, post) \\
PostObligation_a &\equiv I_a(post) \wedge P_a(pre, post)
\end{aligned}$$

The localized verification technique consists of two parts: local correctness and noninterference. Local correctness is shown for each cell separately and completely independently of how the cell is used. Noninterference shows that a cell instantiation does not interfere with the local correctness of the instantiating cell and the instantiated cell.

Local Correctness

All transitions of a cell instance, a, must maintain the invariant and protocol of the cell instance:

$$\textbf{LC} \quad Assumption_a \wedge Transition_a \Rightarrow PostObligation_a$$

By showing this implication for every cell instance of the design, it is verified that all transitions maintain the invariant and protocol locally. But this is not sufficient to guarantee that invariants and protocols of other cell instances are maintained. These can be violated by changing the values of variables passed as parameters.

Noninterference

To show that invariants and protocols involving parameters are preserved, it must be shown that these are not violated by any of the transitions in other cell instances. For example, it must be shown that the transitions of c do not interfere with the invariants and protocols of p, and it must also be shown that the transitions of p do not interfere with the invariants and protocols of c. Instead of showing noninterference for each transition separately, it is done once for each instance of a cell by making the invariants and protocols of a cell instance, c, capture the essential properties of all transitions in a single assertion, $PostObligation_c$. This is expressed in the following two noninterference conditions:

$$\mathbf{NI}\uparrow \quad \mathit{Assumption}_{p,c} \wedge \mathit{PostObligation}_{c} \Rightarrow \mathit{PostObligation}_{p}$$

$$\mathbf{NI}\downarrow \quad \mathit{Assumption}_{p,c} \wedge \mathit{PostObligation}_{p} \Rightarrow \mathit{PostObligation}_{c}$$

Informally, **NI↑** is used to conclude that no transition of the child c interferes with the invariant and protocol of the parent, p, and **NI↓** to conclude that transitions of the parent do not interfere with the protocol and invariant of the child.

Even though many cell instances may share the same state variables, it is only necessary to consider possible interferences between neighbor instances in the instance tree. By showing **NI↑** and **NI↓**, it is also shown that changes made in accordance with the protocol for one neighboring cell instance are in accordance with the protocols for all other neighbors. Therefore, it circumvents the need to consider separately interferences that might result when parameters are passed through several levels, e.g., when a grandchild of a cell instance can change a parameter supplied by the grandparent. In [51], it is shown that the verification technique is sound.

Example: The Iterative Ring

Consider again the iterative ring discussed in sections 2.8.3 and 3.1. Figure 3.9 shows a design description where protocols and parameter restrictions have been added, and these are needed to use the localized verification technique. The instance tree for the design contains one instance of the cell *ring*, which is the root node. The ring instance has n subcell instances of the cell *stage*. The instance verification consists of showing local correctness for each of the $n+1$ cell instances and showing noninterference for the n edges in the instance tree (one for each of the n instantiations of *stage* from *ring*).

To show local correctness of an instance: $\mathit{stage}(v[k], v[k \oplus 1], v[k \oplus 2])$, it must be shown that the transition meets the protocol. This follows directly from the assumption that F preserves the validity of its argument, i.e. that $F(\mathit{empty}) = \mathit{empty}$ and $F(\mathit{valid}) = \mathit{valid}$. The instance of *ring* does not have any transitions, and therefore it follows trivially that it is locally correct.

To verify noninterference, the two steps, **NI↑** and **NI↓**, must be verified for each of the n edges in the instance tree, i.e., for each instantiation of the cell *stage* from the cell *ring*. Substituting the invariant and protocols into the noninterference conditions gives (for typographical reasons the subscript stage_j

```
CELL ring(STATIC n: INTEGER)
  STATIC    E = 0    (* empty value *)
  TYPE
    link = INTEGER
    range = [0..n-1]
    vector = ARRAY range OF link
  STATE   v: vector  RESTRICTION   v: INTERNAL
  FUNCTION valid(l: link): BOOLEAN  RETURN (l <> E)
  FUNCTION drop(STATIC i: INTEGER): BOOLEAN
     RETURN valid(v[i])  AND  NOT valid(v[i⊕1])
  INVARIANT
     FORALL i,j: range. ( drop(i)  AND  drop(j) => (i=j) )   AND
     { OR  i: range | drop(i) }
  CELL stage(in, out, succ: link)
    RESTRICTION
      in, succ: EXTERNAL
      out: LOCAL
    PROTOCOL
      out.pre ≠ out.post =>
      ((out.post=E)=(in.pre=E))  AND  ((out.post=E) ≠ (succ.pre=E))
  BEGIN
    ≪ (in=E) ≠ (succ=E)  ->  out:= F(in) ≫
  END stage
 BEGIN
   {|| k: range | stage(v[k],  v[k⊕1],  v[k⊕2]) }
 END ring
                                    i⊕k is a shorthand for (i+k) MOD n
```

Figure 3.9 Iterative ring

is used as a shorthand for $stage(j, j \oplus 1, j \oplus 2)$):

NI↑ $I_{stage_j}(pre) \wedge I_{ring}(pre) \wedge I_{stage_j}(post) \wedge P_{stage_j}(pre, post)$
$\Rightarrow I_{ring}(post) \wedge P_{ring}(pre, post)$

NI↓ $I_{stage_j}(pre) \wedge I_{ring}(pre) \wedge I_{ring}(post) \wedge P_{ring}(pre, post)$
$\Rightarrow I_{stage_j}(post) \wedge P_{stage_j}(pre, post)$

As I_{stage_j} and P_{ring} are implicitly true, the above reduces to:

NI′↑ $I_{ring}(pre) \wedge P_{stage_j}(pre, post) \Rightarrow I_{ring}(post)$

NI′↓ $I_{ring}(pre) \wedge I_{ring}(post) \Rightarrow P_{stage_j}(pre, post)$

In the first implication (**NI′↑**), the protocol $P_{stage_j}(pre, post)$ is used to conclude that the invariant is maintained for the state variables adjacent to the changing state variable. From $I_{ring}(pre)$ it follows that the invariant is maintained elsewhere. The second implication (**NI′↓**) follows from the restrictions on formal parameters. The state variable $v[k \oplus 1]$ (corresponding to out) can only be changed by stage k.

In principle, **NI↑** and **NI↓** must be shown for each k in the range $0..n-1$. However, the argument is the same for all values of k. It is, therefore, sufficient to show the verification conditions once leaving k as a free variable. This is the basis of the syntax directed verification presented in the next section.

3.4.2 Syntax Directed Verification

The syntax directed verification is very similar to the instance oriented verification; it consists of showing local correctness and noninterference, but it requires fewer verification conditions than the instance oriented approach.

The instance oriented formulation above has a local correctness condition for each instance of a cell. However, the correctness condition is (almost) independent of how the cell is instantiated, and therefore the syntax directed approach merges the local correctness conditions to one generic condition for each cell.

Similarly, it is not necessary to show the two noninterference conditions, **NI↑** and **NI↓**, for each edge in the instance tree, because there are many similar edges in the instance tree stemming from the same generic cell instantiation in the design. For instance, if p and c are instances of the cells P and C, then the edge from p to c in the instance tree is produced by the instantiation of C from P. If there are other instances p' of P, the same instantiation of C from P would also cause another edge from p' to c'. Whereas the instance oriented formulation of **NI**$\uparrow$ and **NI**$\downarrow$ has separate noninterference conditions for (p, c) and (p', c') the syntax directed formulation merges these two similar set of conditions into one set of generic noninterference conditions.

The gain is a reduction in the number of verification condition into a number that only grows linearly with the size of the (syntactic) design description, and independently of the design parameters (which determine the size of the instance tree, i.e., the size of the design).

The localized verification technique makes it possible to verify modular designs by localized arguments involving at most two neighboring levels of the cell hierarchy. By dividing the verification into a number of independent localized conditions, it is possible to verify properties of an incomplete design where some parts are not yet fully designed. The iterating ring shows how it is possible to verify an invariant about the interface between a number of cells without considering the internal details of the cells. The localized verification technique is used by ST2LP for generating verification conditions. This is illustrated in section 3.4.2, where it is shown how the recursively defined arbiter (see section 2.9), is verified.

Example: A Modular Traffic Light Controller

The modular verification technique is illustrated by using it on a modified version of the traffic light controller with two instances of a cell. The modular design description is shown in figure 3.10. The transitions changing the lights in the two directions are similar, and therefore, they are described by a cell which is instantiated twice. This simple example illustrates how to do a localized verification; however, it is too small to exhibit the benefits which are better illustrated by the example in section 3.4.2

The modular verification consists of showing local-correctness of the cell $direction$ and noninterference of each of the two cell instantiations.

```
CELL trafficlight
   STATIC green = 0 yellow = 1 red = 2
   STATE
      EWlight, NSlight: INTEGER
      CarOnEW, CarOnNS: BOOLEAN
   INVARIANT  EWlight=red OR NSlight=red
   CELL direction(thislight, otherlight: INTEGER; sensor: BOOLEAN)
      INVARIANT      thislight=red OR otherlight=red
   BEGIN
      ≪ (thislight = green) AND sensor  ->  thislight:= yellow ≫ ||
      ≪ thislight = yellow  ->  thislight, otherlight:= red, green ≫
   END direction
BEGIN
   direction(EWlight, NSlight, CarOnNS)  ||
   direction(NSlight, EWlight, CarOnEW)
END trafficlight
```

Figure 3.10 Modular traffic light controller

Local Correctness

To show local correctness of the cell *direction*, it must be shown that each of the transition descriptions in this cell meet the verification condition.

$$\textbf{LC} \quad Assumption_a \wedge Transition_a \quad \Rightarrow PostObligation_a$$

In this case, $Assumption_a \wedge PostObligation_a$ degenerates into the invariant of the cell *direction*. Hence, the verification condition for the first transition description becomes:

$$\begin{array}{l}(thislight.pre = red\ OR\ otherlight.pre = red)\ AND \\ (thislight.pre = green)\ AND\ sensor.pre\ AND\ (thislight.post = yellow)\ AND \\ (otherlight.pre = otherlight.post)\ AND\ (sensor.pre = sensor.post) \\ \quad \Rightarrow (thislight.post = red\ OR\ otherlight.post = red)\end{array}$$

A similar verification is needed for the second transition.

Noninterference

There are two cell instantiations and for each there are two noninterference conditions, **NI↑** and **NI↓**, which must be verified.

$$\textbf{NI}\uparrow \quad Assumption_{p,c} \wedge PostObligation_c \quad \Rightarrow PostObligation_p$$

$$\textbf{NI}\downarrow \quad Assumption_{p,c} \wedge PostObligation_p \quad \Rightarrow PostObligation_c$$

In the traffic light controller there are no protocols and substituting the relevant invariants into the conditions for the first instantiation, the various predicates become:

$$\begin{array}{rl}Assumption_{p,c} \equiv & (EWlight.pre = red\ OR\ NSlight.pre = red)\ AND \\ & (EWlight.pre = red\ OR\ NSlight.pre = red)\end{array}$$

The last conjunct is the invariant of the cell *direction*, where the names of the actual parameters are substituted for the corresponding formal parameters.

$$PostObligation_c \equiv EWlight.post = red\ OR\ NSlight.post = red$$

Again the actual parameters have been substituted for the formals.

$$PostObligation_p \equiv EWlight.post = red\ OR\ NSlight.post = red$$

Combining all these, the verification conditions become:

$$\textbf{NI}\uparrow \quad \begin{aligned}&(EWlight.pre = red\ OR\ NSlight.pre = red)\ AND\\ &(EWlight.pre = red\ OR\ NSlight.pre = red)\ AND\\ &(EWlight.post = red\ OR\ NSlight.post = red)\\ &\quad \Rightarrow (EWlight.post = red\ OR\ NSlight.post)\end{aligned}$$

$$\textbf{NI}\downarrow \quad \begin{aligned}&(EWlight.pre = red\ OR\ NSlight.pre = red)\ AND\\ &(EWlight.pre = red\ OR\ NSlight.pre = red)\ AND\\ &(EWlight.post = red\ OR\ NSlight.post = red)\\ &\quad \Rightarrow (EWlight.post = red\ OR\ NSlight.post)\end{aligned}$$

Two similar conditions are needed to show that second instantiation meets the noninterference condition. In this simple case, the conditions degenerate and become trivial, which is not the case for more complicated examples. Exercise 3.9 covers a variation of the traffic light controller where the invariants of the the parent and child cells are different.

Example: Verifying the Arbiter

Consider once more the hierarchical arbiter that was introduced in section 2.9. The localized verification technique can be used to show that no node of the arbiters grants access to both of its children simultaneously, and that all request/grant pairs follow the four-phase protocol. Figure 3.11 shows a design description of the arbiter including the invariant and protocol.

A cell of the arbiter contains five transitions and two cell instantiations. Local correctness must be shown for each of the five transitions, and there are therefore five local correctness conditions to show. For each of the two cell instantiations, two steps are needed to show noninterference (**NI**↑ and **NI**↓). This yields four noninterference conditions. In total there are nine verification conditions. Note that *the number of verification conditions is independent of the depth of the arbiter tree.*

It is possible to mechanically verify the design description shown in figure 3.11 using the translator ST2LP and LP. This involves showing the nine verification conditions mentioned above, and the total verification time is around 1 minute on a standard workstation.

```
CELL arbiter(grp, reqp: BOOLEAN; STATIC depth: INTEGER)
 STATE grl, reql, grr, reqr: BOOLEAN
 RESTRICTION
   grp: EXTERNAL
   reqp, grl, grr: LOCAL
   reql, reqr: INTERNAL
 INVARIANT
   NOT(grl AND grr) AND ( (grl OR grr ) => (grp AND reqp) )
  PROTOCOL
   ((grl.pre AND NOT grl.post) => (NOT reql.pre AND NOT reql.post)) AND
   ((grr.pre AND NOT grr.post) => (NOT reqr.pre AND NOT reqr.post))
BEGIN
 (* requestparent *)
  ≪ NOT grp AND (reql OR reqr) -> reqp:= TRUE ≫ ||
 (* grantleft *)
  ≪ grp AND reqp AND reql AND NOT grr -> grl:= TRUE ≫ ||
 (* grantright *)
  ≪grp AND reqp AND reqr AND NOT grl AND NOT reql -> grr:=TRUE≫||
 (* doneleft *)
  ≪ NOT reql AND grl -> grl, grr, reqp:= FALSE, reqr, reqr ≫ ||
 (* doneright *)
  ≪ NOT reqr AND grr -> grr, reqp:= FALSE, FALSE ≫ ||
 (* instantiate children *)
  {depth > 0 | arbiter(grl, reql, depth-1) || arbiter(grr, reqr, depth-1) }
END arbiter
```

Figure 3.11 Hierarchical arbiter

3.5 BACKGROUND

There are many attempts to provide hardware designers with mechanical verification tools. Most of these aim at implementation verification. i.e., verifying that one circuit description is a refinement of another. Typically, two circuit descriptions are manipulated and transformed into one or more logical expressions, which are then handled by either a theorem prover or a model checker. The survey paper [32] gives a very comprehensive overview of this area. This chapter aims at a somewhat different type of verification: design verification, where one verifies a certain property of a design. Verification of invariance using induction was proposed by Floyd [22]. There are several formalizations based on pre- and post-states using a variety of different notations to distinguish these, however, all of these treat it merely as a notational convention.

Many theorem provers are currently under development in various research centers. The most well known are probably the Boyer-Moore prover [4, 5] and HOL (for hardware verification) [29, 30]. There are many technical differences between these theorem provers, for example, the use of heuristics to formulate additional conjectures in a search for a proof found in Boyer-Moore, and the possibility of users to define their own proof tactics found in HOL and Isabelle [55]. It is beyond the scope of this book to discuss the use of other theorem provers, but it is likely that the mechanical verification of designs in Synchronized Transitions could also be done using one of the other available provers.

The properties verified in this book are all examples of safety properties; they express that "bad things are not happening," for example, that the computation does not enter states violating the invariant. Much current research is directed towards techniques for verifying liveness properties, that is, for proving that certain events *must eventually* occur [20]. Such formal liveness proofs are not considered in this book. The most important reason is to simplify the verification procedure. Formal and mechanized verification of safety properties is already rather difficult. In fact it is so difficult that more work is needed to demonstrate its applicability beyond small examples such as the ones shown in this book. Furthermore, when using mechanical theorem provers, it must always be kept in mind that one is up against undecidable properties and combinatorial explosions. Therefore, it cannot be expected that a mechanical tool can support verification conditions in very sophisticated logics.

3.6 EXERCISES

Exercise 3.1 State a protocol expressing the constraint that the signal $EWlight$ always follow the cycle: $green, yellow, red, green, \ldots$ Can this constraint also be expressed as an invariant?

Exercise 3.2 Is the assertion:

$$NOT((EWlight = green)\ AND\ (NSlight = green))$$

an invariant of the traffic light controller described in section 3.2?

Exercise 3.3 Consider the invariant for the problem of removing duplicate characters given in section 3.1. Use the correctness condition **C** to verify the invariant.

Exercise 3.4 Consider the problem of removing duplicate characters presented in section 3.1. Formulate an assertion stating that the number of character changes before the end markers remains constant. A character change is defined as two different characters next to each other. Verify that the formulated assertion is an invariant for the design given in section 3.1.

Exercise 3.5 Consider again the toggle design from exercise 2.11. Find and verify an invariant ensuring that x and y are never true simultaneously.

Exercise 3.6 Find an invariant for the solution to exercise 2.3 and show that your design maintains it.

Exercise 3.7 Find an invariant for the solution to exercise 2.6 and show that your design maintains it.

Exercise 3.8 Consider the design description shown in figure 3.12. Use the localized verification technique described in section 3.4.1 to:

1. List all the verification conditions needed to show local correctness.
2. List all the verification conditions needed to shown noninterference.

```
CELL fifo
  STATE a, b, c, d: INTEGER   RESTRICTION a, b, c: INTERNAL d: LOCAL
  FUNCTION I2(x, y: INTEGER): BOOLEAN
     RETURN ((x=E)=(y=E)) <=> (x = y)
  CELL copy(x, y, next: INTEGER)
    RESTRICTION x, next: EXTERNAL y: LOCAL
    INVARIANT   I2(x, y) AND I2(y, next)
  BEGIN
    ≪ (x=E) ≠ (next=E) -> y:= x ≫
  END copy
  INVARIANT  I2(a, b) AND I2(b, c) AND I2(c, d)
BEGIN
  copy(a, b, c) || copy(b, c, d) || ≪ d:= c ≫
END fifo
```

Figure 3.12 An example of a modular design

Exercise 3.9 Consider a variation of the modular traffic light controller described in section 3.4.2 where the invariants in the main cell and in the cell *direction* are different.

I_1: *EWlight=red OR NSlight=red*
I_2: *NOT ((EWlight=green) AND (NSlight=green))*

Give the noninterference conditions for the instantiation;

direction(EWlight, NSlight, CarOnNS)

in each of the cases:

1. I_1 is the invariant of the main cell and I_2 is the invariant of the cell *direction*,
2. I_2 is the invariant of the main cell and I_1 is the invariant of the cell *direction*.

Exercise 3.10 Give the five local correctness conditions needed to show local correctness of the arbiter presented in section 3.4.2

4
SYNCHRONOUS DESIGNS

The designs presented in the first three chapters have all been asynchronous. The nondeterminism inherent in these asynchronous computations often leads to simple and succinct descriptions. However, there are also examples where the asynchronous approach is awkward, e.g., when describing computations where a number of components operate in lock step. This chapter shows how to describe such computations as synchronous designs. Syntactically, these designs are similar to asynchronous designs, and most of the constructs introduced in chapters 1 and 2 can also be used for describing synchronous designs, for example, transitions, cells, and quantification. However, the computational model of a synchronous design is different; therefore, a different verification technique is required.

4.1 THE SYNCHRONOUS COMBINATOR

This section introduces the synchronous combinator, $+$. It plays a role similar to the asynchronous combinator, $\|$. To motivate the use of the synchronous combinator, consider a simple shift register, for shifting n-bit vectors.

$$\mathit{TYPE}\ \ vector = \mathit{ARRAY}\,[1..n]\ \mathit{OF\ BOOLEAN}$$
$$\mathit{STATE}\ \ v : vector$$

The bit vector, v, must act like a shift register, i.e., $v[2]$ gets the value of $v[1]$, $v[3]$ the value of $v[2]$, etc. How is such a shift register described in Synchronized Transitions? One may suggest the following design, but it is not correct:

$$\{\ \|\ i : [1..n-1]\ |\ \ll v[i+1] := v[i] \gg\ \}$$

Why not? It is because the $n-2$ transitions are composed asynchronously, and therefore, they can be executed in any order, for example, the following:

$$v[2] := v[1] \;\; v[3] := v[2] \;\; \ldots v[n] := v[n-1]$$

After executing this sequence, all n elements of the vector contain the same value, which was not the intention. Alternatively, one could describe the shift register as follows:

$$\ll v[2], v[3], \ldots, v[n] := v[1], v[2], \ldots, v[n-1] \gg$$

This describes the correct computation, because the value of $v[1]$ is assigned to $v[2]$, $v[2]$ to $v[3]$, etc; but syntactically, it is clumsy and not easy to generalize to an arbitrary value of n. A better solution is obtained by using the the synchronous combinator, $+$, that describes computations where a number of transitions are executed simultaneously, for example:

$$\begin{array}{l} \ll v[2] := v[1] \gg + \\ \ll v[3] := v[2] \gg + \\ \quad \ldots \\ \ll v[n] := v[n-1] \gg \end{array}$$

This describes a computation where the $n-1$ values, contained in $v[1], v[2], \ldots,$ $v[n-1]$ are simultaneously transferred to $v[2], v[3], \ldots, v[n]$. In general, all the expressions on the right-hand-side expressions of all transitions are evaluated, and these values are assigned to the state variables appearing on the left-hand side of the assignments.

Syntactically, the synchronous combinator plays exactly the same role as the asynchronous combinator $||$. It may, for example, be used in a quantified instantiation such as the following:

$$\{+\; i : [1..n-1] \mid \ll v[i+1] := v[i] \gg \}$$

This is a compact description of the shift register which can be used for any value of n. This is not the case with the multi-assignment shown above.

Exclusive Write

To avoid conflicting assignments, synchronous composition is only defined if the variables on the left-hand sides of all assignments are distinct. This is expressed more precisely by the exclusive write condition given below.

To compose a set of transitions, $t_1, t_2, \ldots, t_n$, synchronously, the set must meet the exclusive write condition (EW).

Definition 4.1 *The transitions $t_1, t_2, \ldots, t_n$ meet the Exclusive Write Condition if and only if:*

$$\forall i,j \in [1..n] : W^{t_i} \cap W^{t_j} \neq \emptyset \Rightarrow \neg(c_i \wedge c_j)$$

W^{t_i} is the write set of transition t_i, i.e., the set of variables appearing on the left-hand side of its multi-assignment (see appendix A section A.14.1). The exclusive write condition requires that transitions with intersecting write sets are not enabled simultaneously.

The shift register meets the exclusive write condition, because all the variables appearing on the left-hand sides are distinct, which means that all write sets are disjoint ($W^{t_i} \cap W^{t_j} = \emptyset$). In more complicated cases, it is often necessary to utilize invariants of the design to show that the exclusive write condition is met.

4.1.1 Computational Model

To define the computation of a synchronous design, consider a collection of n transition descriptions, t_i, $(0 < i \leq n)$ that meet the exclusive write condition:

TRANSITION $t_i \ll c_i \texttt{->} v_i := e_i \gg$

The synchronous composition of these n transitions: $t_1 + t_2 + \ldots + t_n$ describes an iterative computation where each iteration consists of two phases. During the first phase, all the n preconditions, c_i, and right-hand side expressions, e_i, are evaluated. In the second phase, all the variables, v_i, on the left-hand side of the enabled transitions are updated with the new values specified by the corresponding expressions e_i. The preconditions and the right-hand side expressions are evaluated *before* any of the state variables are changed. Hence, it is the values of the state variables in the pre-state which are used.

The iterative nature of the synchronous model corresponds to repeated selection in the the model of asynchronous designs. However, in the synchronous model there is no nondeterministic selection of the transition to execute, instead *all* enabled transitions are executed. In particular, transitions that are always enabled, e.g., because their preconditions are $TRUE$, will be executed in every iteration. Synchronous designs are used to model synchronous circuits controlled by a global clock, and each iteration corresponds to a clock cycle. In chapter 5 it is illustrated how a synchronous design is transformed into a corresponding synchronous circuit.

From a theoretical point of view, synchronous composition does not add any additional expressive power. It is possible to translate any design description using synchronous composition into an equivalent form using asynchronous composition only. Consider, for example, the synchronous composition of the three transitions: $t_1 + t_2 + t_3$. The following asynchronous design describes the same computation:

$$
\begin{array}{llllll}
\ll c_1 & \mathit{AND}\ c_2 & \mathit{AND}\ c_3 & \text{->} & v_1, v_2, v_3 := e_1, e_2, e_3 & \gg \| \\
\ll c_1 & \mathit{AND}\ c_2 & \mathit{AND\ NOT} c_3 & \text{->} & v_1, v_2 := e_1, e_2 & \gg \| \\
\ll c_1 & \mathit{AND\ NOT} c_2 & \mathit{AND}\ c_3 & \text{->} & v_1, v_3 := e_1, e_3 & \gg \| \\
\ll c_1 & \mathit{AND\ NOT} c_2 & \mathit{AND\ NOT} c_3 & \text{->} & v_1 := e_1 & \gg \| \\
\ll \mathit{NOT} c_1 & \mathit{AND}\ c_2 & \mathit{AND}\ c_3 & \text{->} & v_2, v_3 := e_2, e_3 & \gg \| \\
 & & \dots & & & \\
\ll \mathit{NOT} c_1 & \mathit{AND\ NOT} c_2 & \mathit{AND\ NOT} c_3 \gg & & &
\end{array}
$$

This way of describing a computation is clumsy and impractical for larger designs. So although the asynchronous combinator is sufficient in theory, the synchronous combinator is needed to make practical designs tractable.

Mixing the Combinators

All the examples discussed in this chapter use the synchronous combinator only. The language report in appendix A states that the combinators cannot be mixed and almost all designs in this book are either purely asynchronous or purely synchronous. The only exceptions are in section 8.1 that describes some designs where the combinators are mixed.

Example: A Taxameter

This section describes the design of a simple taxameter with the incoming signals: *on*, *sec*, *m100*, and *km*. When the taxameter is switched on, by making *on* true, it must generate a *clear* signal that goes to a (mechanical) display, and sets it to zero. While the taxameter is on, it must generate a *count* signal for each minute, and for each *m100* signal received. The signal *m100* comes from the odometer every time the taxi has travelled 100 meters. When the signal *km* is received (from a push-button) the taxameter must generate a *show* signal and hold it for 10 seconds. This signal makes the display show the accumulated distance to the driver. The *sec* signal comes from an external clock once every second. The interface to the taxameter is shown below.

```
STATIC  secpmin = 60
CELL minclock(minute, sec: BOOLEAN)
  RESTRICTION minute: LOCAL sec: EXTERNAL
  TYPE range = [1..secpmin]
  STATE minc: range
  INITIALLY minc = secpmin
BEGIN
  ≪ minute  –>  minute, minc:= FALSE, 1 ≫ +
  ≪ (minc < secpmin) AND sec  –>  minc:= minc+1 ≫ +
  ≪ NOT minute AND (minc ≥ secpmin)  –>  minute:= TRUE ≫
END minclock
```

Figure 4.1 Clock generating signals every minute

```
STATIC showlim = 10
CELL display(show, km, sec: BOOLEAN)
  RESTRICTION show: LOCAL km, sec: EXTERNAL
  TYPE range = [1..showlim]
  STATE showc: range
  INITIALLY showc = showlim
BEGIN
  ≪ km AND NOT show  –>  show, showc:= TRUE, 1 ≫ +
  ≪ (showc < showlim) AND sec  –>  showc:= showc+1 ≫ +
  ≪ show AND (showc ≥ showlim)  –>  show:= FALSE ≫
END display
```

Figure 4.2 Display generating the signal show

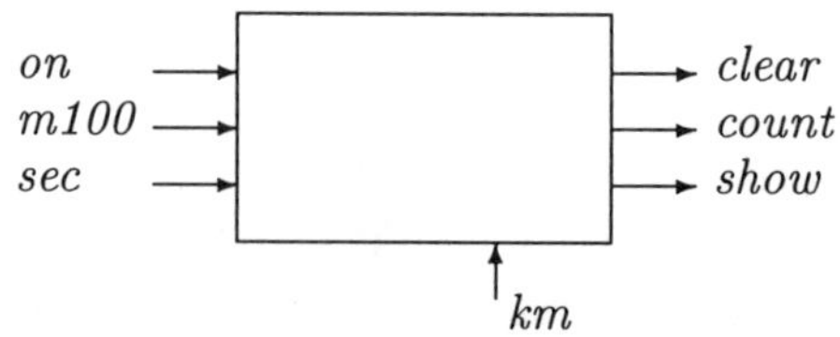

The incoming signals, for example *on*, are assumed to retain their value during an entire iteration (clock cycle), but they may change from one iteration to the next. The signal *sec* is true for one iteration with a period of one second. The outgoing signals, for example *count*, must follow the same convention, i.e., it is true for one iteration every time the display is incremented. If it is true in two consecutive iterations, this is interpreted as two separate increments.

The design description of the taxameter is divided into three cells:

- *minclock*, that generates a signal every minute,
- *display*, that keeps track of the display, and
- *counting*, that generates the clear and count signals.

Figure 4.1 shows the the cell *minclock* where the input signal *sec* is used to generate the signal *minute*. In this cell, the local variable *minc* is incremented every time *sec* is true. When *minc* reaches the value *secpmin* (60), the state variable *minute* is true for one iteration.

The first and second transitions both write the state variable *minc*, and the first and third both write *minute*. It is quite simple to see that the first and third transitions meet the exclusive write condition, because *minute* and *NOT minute* cannot be true simultaneously. But it is less obvious that the first and second transitions also meet the condition, i.e., that the preconditions of these transitions cannot both hold. This is, however, the case; one way to see this is to realize that:

$$minute \Rightarrow minc \geq secpmin$$

is an invariant for the cell *minclock*. From this invariant, it follows that the preconditions of the first and second transitions never hold simultaneously; hence, exclusive write is met. Section 4.2 describes a technique for verifying that a synchronous design satisfies an invariant.

The cell *display*, shown in figure 4.2, is very similar to *minclock*. It ensures that *show* is true for *showlim* seconds. The reasoning to show that the cell *display*

```
CELL taxameter(sec, show, on, clear, m100, km, count: BOOLEAN)
  RESTRICTION
     sec, on, m100, km: EXTERNAL
     show, clear, count:INTERNAL
  STATE  active: BOOLEAN   RESTRICTION active: INTERNAL
  CELL counting(sec, on, clear, m100, count, active: BOOLEAN)
     RESTRICTION
       sec, on, m100: EXTERNAL
       clear, count, active: LOCAL
     STATE minute: BOOLEAN      RESTRICTION minute: INTERNAL
     INITIALLY minute = FALSE  active = FALSE
     CELL minclock(minute, sec: BOOLEAN)
       ...
     END minclock
  BEGIN
     minclock(minute, sec) +
     ≪ NOT on AND active AND NOT clear -> active:= FALSE ≫  +
     ≪ on AND NOT active -> active, clear:= TRUE, TRUE ≫  +
     ≪ clear -> clear:= FALSE ≫  +
     ≪ active AND on AND (m100 OR minute) -> count:= TRUE ≫  +
     ≪ (NOT (m100 OR minute) OR NOT on) -> count:= FALSE ≫
  END counting
  CELL display(show, km, sec: BOOLEAN)
    ...
  END display
BEGIN
  counting(sec, on, clear, m100, count, active)  +  display(show, km, sec)
END taxameter
```

Figure 4.3 Taxameter design

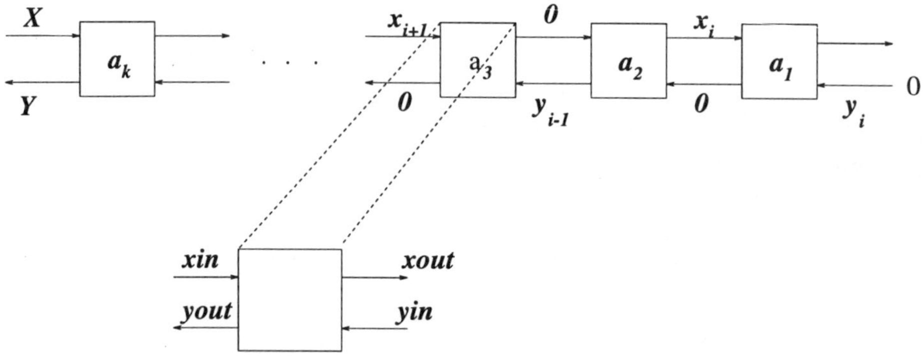

Figure 4.4 Structure of FIR filter

meets the exclusive write condition is like the one used for the cell *minclock* (see exercise 4.5). Figure 4.3 shows a design description of the taxameter where the two cells *minclock* and *display* are combined with a number of transitions describing some control logic (the internal details of these two cells is omitted). This cell also illustrates how cells can be composed with the synchronous combinator.

Example: A Systolic FIR filter

Systolic designs are synchronous computations with a regular structure that makes it possible to describe them by repeated instantiation of the same cell. Data move between these cells and are processed in a fixed number of iterations. Systolic designs rely on the lock step nature of synchronous computations. This makes it possible to control exactly where the moving data elements are located in each iteration.

The FIR (Finite Impulse Response) filter is a simple, but typical, example of a systolic design. It produces a sequence of output data, $Y = y_1, y_2, \ldots, y_i, \ldots$ from a sequence of input data, $X = x_1, x_2, \ldots, x_i, \ldots$ by computing an inner product of the form:

$$y_i = \sum_{j=1}^{j=k} a_j \cdot x_{i+j-1}$$

The order k is fixed. The filter computation is described as a linear array of simple processing elements (see figure 4.4). In every other iteration, a new value of *X* is inserted in the array (at the point indicated with *X* in figure 4.4).

In the remaining iterations the value 0 is inserted. Similarly, a new element of the output sequence, Y, is computed in every other iteration (the output appears at the point indicated with Y in figure 4.4). At the point indicated with 0 the value *0* is input in every iteration, and all cells compute a multiplication and accumulate the result. Furthermore, the x-value is propagated which makes the sequence of inputs, X, move from left to right.

```
CELL multacc(xin, xout, yin, yout : element)
  STATE a : element
BEGIN
  ≪ yout := yin + a * xin ≫ +
  ≪ xout := xin ≫
END multacc
```

Each cell receives an element of X and a partially computed element from Y. It propagates the x-value to the right and the updated y-value to the left. Figure 4.5 shows the complete design of the filter parameterized by the value of k. This design does not specify how the filter coefficients (a) are supplied (see exercise 4.7) or how the input/output of the sequences is done.

The taxameter and filter show how synchronous computations are described using the synchronous combinator. In the taxameter, it is essential to keep the counting synchronized with the control logic, and in the filter data must meet at the processing elements at exactly the correct iteration of the computation. These examples capture the essence of a synchronous computation where the functional behavior relies on exact timing. Generally, it is quite complicated to design such computations because timing and functionality are mixed. On the other hand, once a correct design has been made, it corresponds directly to a synchronous circuit. Chapter 5 shows how to synthesize clocked circuits from synchronous design descriptions.

Example: The Product Combinator

This section introduces a third combinator, called the product combinator "*". Unlike the synchronous and asynchronous combinators it does not model a particular kind of computation; it is used for structuring a design.

The product combinator is used to factor a transition into a number of simpler transitions. Let t_1, t_2 be the following two transitions that are assumed to meet the exclusive write condition:

```
TRANSITION t1 ≪ c1 -> v1 := e1 ≫
TRANSITION t2 ≪ c2 -> v2 := e2 ≫
```

```
CELL filter(X, Y: element; STATIC k: INTEGER)
  TYPE range = [1..k-1]
  STATE
    x, y: ARRAY range OF element
    zero, dummy: element
  INITIALLY zero = 0
  CELL multacc(xin, xout, yin, yout: element)
      STATE a: element
  BEGIN
     ≪ yout:= yin + a*xin ≫ +
     ≪ xout:= xin ≫
  END multacc
BEGIN
  multacc(X, x[k-1], y[k-1], Y) +
  {+ i: [2..k-1] | multacc(x[i], x[i-1], y[i-1], y[i]) } +
  multacc(x[1], dummy, zero, y[1])
END filter
```

Figure 4.5 Description of the FIR filter design

The product, $t_1 * t_2$, is equivalent to the following transition:

$$\ll c_1 \ AND \ c_2 \ {->} \ v_1, v_2 := e_1, e_2 \gg$$

The product combinator makes it possible to separate transitions dealing with different aspects. As is the case with the synchronous combinator, the product combinator does not add expressive power, but in many cases it yields simpler and more comprehensible designs.

As an important special case, consider a transition description consisting of a precondition only, for example: $\ll active\ AND\ enabled \gg$. The product combinator can be used to factor this precondition out of a number of other transition descriptions. As an example, consider the following transitions:

$$\ll active\ AND\ enabled\ {->}\ run := TRUE \gg +$$
$$\ll active\ AND\ enabled\ AND\ NOT clear\ {->}\ c := c+1 \gg +$$
$$\ll active\ AND\ enabled\ AND\ c = limit\ {->}\ stop := TRUE \gg$$

This description is equivalent to:

$$\ll active\ AND\ enabled \gg *$$
$$(\ \ll run := TRUE \gg +$$
$$\ll NOT clear\ {->}\ c := c+1 \gg +$$
$$\ll c = limit\ {->}\ stop := TRUE \gg\ )$$

The product combinator can also be used to factor an asynchronous design.

Example: A Taxameter with Reset

Consider a variation of the taxameter design where an explicit reset signal is used to force internal state variables to a particular value. This cannot be done with an initialization of the variables using *INITIALLY* which only sets the value of the variables once before the computation starts. In this design, it must be possible to force the variables to the desired values many times during a computation. The reset is an additional external signal to the taxameter.

CELL $taxameter(sec, show, on, clear, m100, km, count, reset : BOOLEAN)$

When *reset* is true, the internal variable *minute* must become false, i.e.,

$$\ll reset\ {->}\ minute := FALSE \gg$$

Similarly, the other internal variables are given a new value when *reset* is true. However, the other transitions manipulating *minute* must also be changed to avoid conflicts when *reset* is true. In particular, no transition writing *minute*

```
CELL minclock(minute, sec, reset: BOOLEAN)
  RESTRICTION minute: LOCAL    sec, reset: EXTERNAL
  TYPE range = [1..secpmin]
  STATE minc: range
BEGIN
  ≪ reset -> minute, minc:= FALSE, secpmin ≫ +
  ≪ NOT reset AND minute -> minute,minc:= FALSE, 1 ≫ +
  ≪ NOT reset AND (minc < secpmin) AND sec -> minc:= minc+1 ≫ +
  ≪ NOT reset AND NOTminute AND (minc≥secpmin) -> minute:= TRUE ≫
END minclock
```

Figure 4.6 A design with reset

must be enabled. A simple solution is to change the transitions in the cell *minclock* by adding the conjunct *NOTreset* to all the preconditions as shown in figure 4.6. The first transition gives an initial value to *minute* and *minc* when *reset* is true. However, all the remaining transitions must have the conjunct *NOTreset* added to their precondition, otherwise the exclusive write condition is not met. By using the product combinator to factor out the common conjunct, a more comprehensible design is obtained. This leads to the following transition descriptions:

$$\ll reset -> minute, minc := FALSE, secpmin \gg +$$
$$\ll NOTreset \gg *$$
$$(\ll minute -> minute, minc := FALSE, 1 \gg +$$
$$\ll (minc < secpmin)\ AND\ sec -> minc := minc + 1 \gg +$$
$$\ll NOTminute\ AND\ (minc \geq secpmin) -> minute := TRUE \gg)$$

This is completely equivalent to the first design description; however, the second version is easier to read (and write).

Example: A Stop Watch

This section describes the design of a simple digital stop watch with a three digit, seven-segment display (to read tens of seconds, seconds, and tenths of seconds), and two control buttons, "reset" and "start/stop". When the reset button is pressed the display is cleared. The start/stop button is used to start and stop the clock. The design is driven by an external 1MHz clock signal.

The external clock signal is used to keep track of the time by counting. A transition in a synchronous design is executed once in every clock cycle, so it is very simple to count the number of clock cycles.

$\ll$ *oneMHz* := (*oneMHz* + *1*) *MOD 1000000* $\gg$

The least significant digit of the display must be incremented every 10th of a second, i.e., with a frequency of 10Hz. The state variable *oneMHz* is used to generate a 10Hz signal as follows:

```
CELL generate(tenHz : BOOLEAN)
  STATE oneMHz : INTEGER
  INITIALLY oneMHz = 1   tenHz = FALSE
BEGIN
  << oneMHz := (oneMHz + 1) MOD 1000000 >>  +
  << tenHz := (oneMHz MOD 100000) = 0 >>
END generate
```

The first transition increments the state variable *OneMHz* once in each iteration (clock cycle), i.e., one million times per second. The state variable *tenHz* is therefore true ten times per second, and every time it is true for one iteration.

The start/stop button, represented by the state variable *ss*, is used to toggle a state variable *run*. When *run* is true, a modulo 10 counter, *c*, is active. This counter corresponds to the least significant digit of the stop watch.

```
CELL counter(reset, ss, run : BOOLEAN; c :  digit)
BEGIN
  << reset -> c := 0 >>  +
  << NOT reset >> *
    ( << run -> c := (c+1) MOD 10 >>  +
      << ss -> run := NOTrun >> )
END counter
```

If the transitions of this cell were executed once every tenth of a second, then *c* would represent the least significant digit (tenth of a second) on the display. By instantiating the cell *counter* as follows:

```
counter(reset, ss, run, c) * << tenHz >>
```

the transitions of the cell *counter* are only enabled when *tenHz* is true, i.e., once every tenth of a second. Therefore, *c* represents the least significant digit of the display. From *c* the other digits are generated by counting. Figure 4.7 shows the complete design description of the stop watch. The design of a stop watch was proposed by Michael Fourman, and in [23] he presents a solution

```
CELL stopwatch(reset, ss: BOOLEAN)    (* ss is the start/stop button *)
  TYPE digit = [0..9]
  STATE
    c: digit
    tenHz, run: BOOLEAN
  CELL counter(reset, ss, run: BOOLEAN; c: digit)
  BEGIN
    ≪ reset -> c:= 0 ≫ +
    ≪ NOT reset ≫ *
      ( ≪ run -> c:= (c+1) MOD 10 ≫ + ≪ ss -> run:= NOT run ≫ )
  END counter
  CELL display(reset, run: BOOLEAN; c: digit)
    (* The control of the physical display not included in
     this design. ten, sec, and c are the digits that must be displayed *)
    STATE sec, ten: digit
  BEGIN
    ≪ reset -> sec, ten:= 0,0 ≫ +
    ≪ NOT reset AND run ≫ *
      ( ≪ c=9 -> sec:= (sec+1) MOD 10 ≫ +
        ≪ sec=9 -> ten:= (ten+1) MOD 10 ≫ )
  END display
  CELL generate(tenHz: BOOLEAN)
    STATE oneMHz: INTEGER
    INITIALLY oneMHz = 1   tenHz = FALSE
  BEGIN
    ≪ oneMHz:= (oneMHz+1) MOD 1000000 ≫ +
    ≪ tenHz:= oneMHz MOD 100000 = 0 ≫
  END generate
BEGIN
  ( counter(reset, ss, run, c) + display(reset, run, c) ) * ≪ tenHz ≫ +
  generate(tenHz)
END stopwatch
```

Figure 4.7 Design of a stop watch

using a completely different approach.

4.2 VERIFICATION OF SYNCHRONOUS DESIGNS

Invariants and protocols are useful for verifying synchronous designs also; however, the verification technique used for asynchronous designs is not applicable, because the computational model of a synchronous design is different from the one used for asynchronous designs. For asynchronous designs, it is possible to verify invariants and protocols by considering one transition at a time. This is not adequate for synchronous designs, as illustrated by the following simple example:

$$\ll a := a{+}1 \gg \quad + \quad \ll b := b{+}1 \gg$$

This synchronous composition maintains the invariant: $a = b$; if this assertion is satisfied before execution of the two transitions, it is also satisfied afterwards, because both a and b are incremented. However, this cannot be verified by considering the two transitions separately. If one of the transitions, for example: $\ll a := a{+}1 \gg$, is considered by itself, it does not maintain the invariant. In this this simple case, the verification can be done by rewriting the two synchronously composed transitions into a single transition.

$$\ll a, b := a{+}1, b{+}1 \gg$$

This rewritten transition does maintain the invariant, and hence, it can be concluded that this is also the case for the synchronous composition of the two original transitions. However, for a large design such a rewriting is not practical, because the resulting transition may become very large, so another approach is needed.

The verification technique for asynchronous design makes it possible to break the verification into into a number of independent steps, one for each transition. In this way, each step of the verification is relatively simple. It would be nice to have a similar technique for synchronous designs. The approach described below allows some amount of separation, but it does not seem possible to get a complete separation, as was the case for asynchronous designs.

When a collection of synchronously composed transitions, t_1, t_2, ..., t_n, are started in a particular pre-state then each of them delivers a small contribution to the post-state. The exclusive write condition ensures that none of the transitions write conflicting values to the same variable. More concretely, let I and P denote the invariant and protocol to be verified, and let Q_i be a predicate describing transition t_i's contribution to the post-state. The first step of the verification must show that the Q_i predicates are consistent with the transition descriptions, which is the case if the following implications can be shown:

$$\forall i \in [1..n] : I(pre) \wedge P(pre, post) \wedge t_i(pre, post) \Rightarrow Q_i(pre, post)$$

Consider again the simple example introduced above, the following two predicates Q_1 and Q_2 describe the contribution of the two transitions:

$$Q_1 : a.post = b.pre + 1 \qquad Q_2 : b.post = a.pre + 1$$

It is simple to show that:

$$(a.pre = b.pre \wedge a.post = a.pre + 1) \Rightarrow a.post = b.pre + 1$$

$$(a.pre = b.pre \wedge b.post = b.pre + 1) \Rightarrow b.post = a.pre + 1$$

Note that this step of the verification is done by considering each transition separately, similarly to the way it is done for asynchronous designs.

The second step of the verification consists of showing that the invariant and protocol hold in the post-state, by using the Q_i predicates. For the simple example, this is straightforward, because:

$$(a.pre = b.pre \wedge a.post = b.pre + 1 \wedge b.post = a.pre + 1) \Rightarrow b.post = a.post$$

thus, it follows that $a = b$ is an invariant.

The example shows how it is possible to break the verification of an invariant into independent steps. However, in addition to the invariants and protocols, the Q_i predicates must be provided which is an additional burden on the designer. Furthermore, the verification technique has a second step where the contributions of all transitions are combined in a single verification condition.

4.2.1 Mechanized Verification of Synchronous Designs

The mechanical tools LP and ST2LP can also be used to verify synchronized designs. The major difference, compared to asynchronous designs, is the set of predicates Q_i which in principle are supplied by the designer. However, in many cases, it is possible for ST2LP to derive these predicates from the transition descriptions, and then it is not necessary to provide them explicitly.

Consider the following transition:

$$\mathit{TRANSITION}\ \ t \ll c \text{->} v := e \gg$$

The following predicate Q describes the contribution of this transition:

$$Q(pre, post) \equiv (c \Rightarrow v.post = e.pre) \wedge (\neg c \Rightarrow v.post = v.pre)$$

As an illustration consider the transition, $\ll a := a + 1 \gg$. From which the following predicate is generated:

$$Q(pre, post) \equiv (TRUE \Rightarrow a.post = a.pre + 1) \wedge (\neg TRUE \Rightarrow a.post = a.pre)$$

It can be simplified to:

$$Q(pre, post) \equiv a.post = a.pre + 1$$

When there are several transitions changing a variable, the derivation of Q is a bit more complicated, but ST2LP handles this.

Once all the Q_i predicates have been established, either explicitly by the designer or automatically by ST2LP, the following implication, called synchronous correctness (SC), must be verified:

$$\textbf{SC} \qquad \bigwedge_{i=1}^{N} Q_i(pre, post) \wedge I(pre) \Rightarrow I(post) \wedge P(pre, post)$$

More details about verification of synchronous designs are given in [51]. The verification of synchronous designs is somewhat more cumbersome than verifying asynchronous designs. The synchronous correctness condition involves all transitions of a design, which on the one hand is unpleasant, because the condition becomes large, but on the other hand, it seems difficult to avoid, because a synchronous design relies on dependencies between the transitions.

Example: Verifying the Stop Watch

Consider again, the generate cell of the stopwatch

CELL $generate(tenHz : $ *BOOLEAN)*
  STATE $oneMHz$: *INTEGER*
  INITIALLY $oneMHz = 0 \quad tenHz = FALSE$
BEGIN
  $\ll oneMHz := (oneMHz + 1)\ MOD\ 1000000 \gg \quad +$
  $\ll tenHz := (oneMHz\ MOD\ 100000) = 0 \gg$
END generate

This cell maintains the invariant:

$$I : tenHz = ((oneMHz\ MOD\ 100000) = 1)$$

It is simple to see that the initial state satisfies the invariant, since both the left and right-hand sides are false. To verify that the transitions maintain invariant, the first step is to derive the Q_i assertions from the two transitions of the cell.

$$Q_1(pre, post) \equiv oneMHz.post = (oneMHz.pre + 1) \; MOD \; 1000000$$
$$Q_2(pre, post) \equiv tenHz.post = (oneMHz.pre \; MOD \; 100000) = 0$$

The second step is to verify synchronous correctness.

$$Q_1(pre, post) \wedge Q_2(pre, post) \wedge I(pre) \Rightarrow I(post)$$

There are two cases to consider.

- $(oneMHz.post \; MOD \; 100000) = 1$,
  in this case $(oneMHz.pre \; MOD \; 100000) = 0$, so $tenHz.post = TRUE$, which means that $I(post)$ holds.
- $(oneMHz.post \; MOD \; 100000) \neq 1$,
  in this case $(oneMHz.pre \; MOD \; 100000) \neq 0$, so $tenHz.post = FALSE$, which means that $I(post)$ holds.

It follows, that I is an invariant for the generate cell.

4.3 A FAST ADDER

This section presents the Brent-Kung adder that illustrates several points. First of all, it is a nontrivial and efficient circuit (for large word lengths there are no faster adders). Second, the adder is complicated, but an invariant is a significant help in understanding how it works. Finally, the Brent-Kung adder is a nice example of a synchronous computation.

To stress the importance of the invariant for understanding the Brent-Kung adder consider the design description shown in figure 4.8. It is not at all obvious that this is a computation of the carry bits needed in an adder. However, once the invariant of the design has been formulated, this is hopefully clear.

The Brent-Kung adder adds n-bit operands in a time proportional to $log\ n$. This is in contrast to a simple ripple carry adder that requires a time proportional to n. For large values of n, it is not possible to add faster than $log\ n$, so speed-wise the Brent-Kung adder is an optimal design.

```
CELL myst(a, b, d: word)
   STATE t: INTEGER
   INITIALLY t = 1
BEGIN
   ≪ t:= 2*t ≫  +
   {+ i: [1..n] | ≪ (i≥t) AND (d[i]=U) -> d[i]:= d[i-t] ≫ }
END myst
```

Figure 4.8 What is this design computing?

4.3.1 Invariant for the Fast Adder

Let a and b be two n-bit vectors that are going to be added. To do this, it is necessary to compute the vector of carry values, called c which a $n+1$-bit vector.

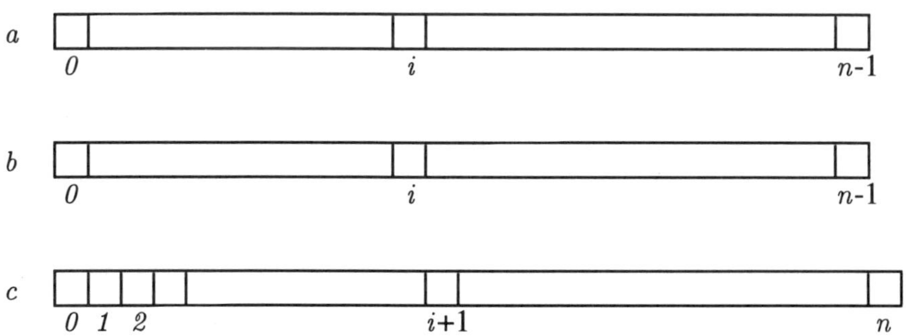

By definition

$$c[i+1] = \begin{cases} 1 & \text{if } a[i] + b[i] + c[i] > 1 \\ 0 & \text{otherwise} \end{cases}$$

Another way of expressing this is:

$$c[i+1] = \begin{cases} a[i] & \text{if } a[i] = b[i] \\ c[i] & \text{if } a[i] \neq b[i] \end{cases}$$

Which can be generalized to:

$$c[i+1] = a[m_i];$$

where m_i is determined by

$$(\forall\, j : m_i < j \leq i : \; a[j] \neq b[j]) \wedge (a[m_i] = b[m_i])$$

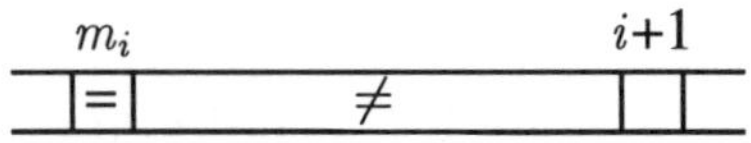

The last expression for the carry, $c[i+1]$, is used to formulate the invariant where c denotes the correct carry that is ultimately computed by the algorithm. To compute c the state variable d is used, during the computation it contains intermediate approximations of the correct carry, and the goal of the carry computation is to establish that:

$$\forall i \in [0..n] : d[i] = c[i]$$

The elements of d take one of three values, 0, 1, or U. The last can be read as "unknown" indicating that the correct carry value has not yet been computed for corresponding element of d. Therefore, all elements of d are initialized with the value U. During the computation, $d[i] = U$ if there is a consecutive sequence of indices of length t, where $a[j] \neq b[j]$:

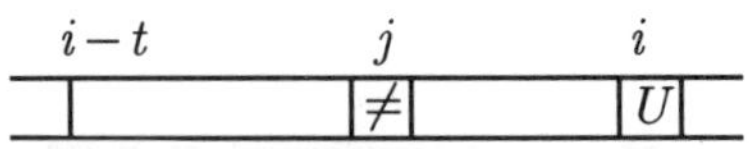

If on the other hand, there is no such sequence of consecutive indices, then $d[i]$ must contain the correct carry, i.e. $c[i]$.

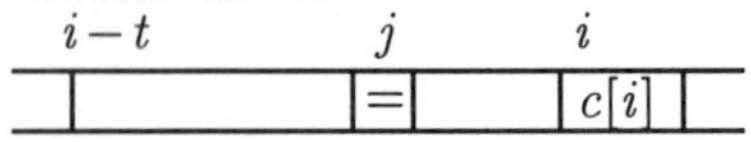

The state variable t is initially 1 and gradually incremented during the computation. The following invariant is a more succinct way of describing the property illustrated by the diagrams shown above.

$$\begin{array}{ll} 0 \leq i < t : d[i] = & c[i] \qquad\qquad\qquad\qquad\qquad\qquad\qquad \wedge \\ t \leq i \leq n : d[i] = & \begin{cases} c[i] & \text{if } \exists j : i-t \leq j < i : a[j] = b[j] \\ U & \text{otherwise} \end{cases} \end{array}$$

The invariant states that $d[i] = U$ only if there is a preceding sequence of consecutive indices of length t where $a[j] \neq b[j]$. This invariant is the key to the construction of the fast adder.

The objective of the carry computation is to eliminate all U values from d. To see how $d[i]$ is updated, consider the following intermediate stage of the computation:

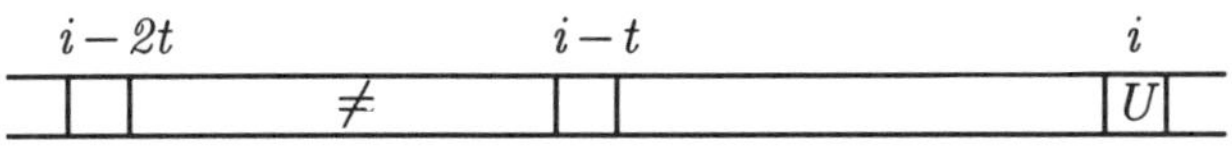

Assume that the invariant holds, then the value of $d[i]$ can be updated by copying the value of $d[i-t]$. To understand why this is correct consider the two cases:

$d[i-t] \neq U$ then the invariant states that $d[i-t]$ contains the correct carry,

$d[i-t] = U$ the invariant tells us that we have the following situation:

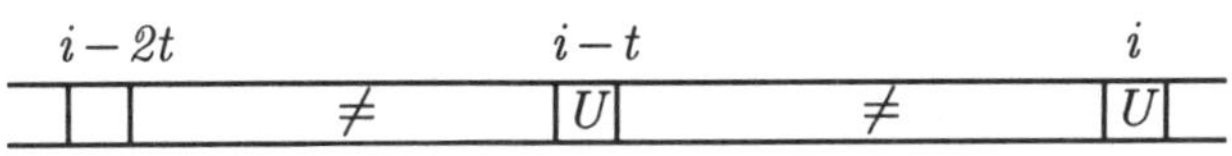

Again it is correct to copy the value of $d[i-t]$ to $d[i]$.

It can therefore be concluded, that the invariant is maintained by the transition.

$$\ll (i \geq t)\ AND\ (d[i] = U) \rightarrow d[i] := d[i-t] \gg$$

However, the argument given above also shows that t can be doubled without invalidating the invariant. It has therefore been shown that the following transitions maintain the invariant

$$\{+\ i:\ [0..n] \mid \ll (i \geq t)\ AND\ (d[i] = U) \rightarrow d[i] := d[i-t] \gg \}\ +$$
$$\ll t := 2 * t \gg$$

These transitions for determining the carry are the key to the fast adder. Note that t is doubled in every iteration while maintaining the invariant. One of the conjuncts of the invariant is:

$$0 \leq i < t : d[i] = c[i]$$

Since t is doubled in every iteration, the correct carry is found in $log\ n$ steps. Once the carry has been found, the sum can be determined in a single step:

$$\{+\ i:\ [0..n-1] \mid \ll s[i] := (a[i] \neq b[i]) \neq c[i] \gg \}$$

The only remaining part of the design is the initialization which must establish the invariant. This is done as follows:

$d[0] = 0$
$\{i : [0..n-1] \mid d[i+1] = \mathit{IF}\ a[i] = b[i]\ \mathit{THEN}\ a[i]\ \mathit{ELSE}\ U\ \}$
$t = 1$

4.3.2 Realization

Figure 4.9 shows a complete design of the fast adder, a very similar design is given by Chandy and Misra in [10], and therefore, this adder is called the Chandy-Misra adder. The Chandy-Misra adder can compute the sum of two n-bit binary words in $\log n$ iterations; however, there are several things in the design that makes it hard to directly derive an efficient integrated circuit realization:

- elements of d can have three different values (0, 1, and U); therefore, it cannot be represented directly as a binary vector,
- the assignments $d[i] := d[i-t]$ require connections between the circuitry realizing the different array elements. Taking all assignments into account leads to a large number of crossing wires which cause problems in an integrated circuit,
- the integer value t must be distributed to a large number of subcircuits and such a broadcasting may slow down the entire circuit.

In the next section, it is shown how the fast adder can be modified to avoid these problems. This modified adder is the Brent-Kung adder [7], which is recognized as a very efficient design.

4.3.3 The Brent-Kung Adder

The fast adder presented in the previous section can be refined so the problems mentioned in section 4.3.2 are avoided. These refinements lead to the Brent-Kung adder.

```
STATIC U = 2
TYPE
   wordrange = [0..n]
   wordval = [0..1]
   word = ARRAY wordrange OF wordval
   carryval = [0..2]  (* 0, 1, U *)
   carryword = ARRAY wordrange OF carryval
STATE
   a, b, s: word
   d: carryword
CELL carry(a, b: word; d: carryword)
   STATE t: INTEGER
   INITIALLY t = 1
   FUNCTION prop(a, b: word; i,t: INTEGER): BOOLEAN
     RETURN {AND j: wordrange | (i-t≤j)  AND  (j<i) => a[j] ≠ b[j] }
   INVARIANT
     FORALL i:  wordrange . (0 < t) AND
        (t > i =>   d[i] = c[i]) AND
        ( t≤i =>  d[i] = IF  prop(a, b, i, t) THEN U ELSE c[i])
BEGIN
   {+ i: [1..n] | ≪ (i≥t)  AND  (d[i]=U)  ->  d[i]:= d[i-t] ≫ }  +
   ≪ t:= 2*t ≫
END carry
INITIALLY
  d[0] = 0
  {i: [0..n-1] | d[i+1] = IF a[i]=b[i] THEN a[i] ELSE U }
BEGIN
  carry(a, b, d) +
  {+ i: [0..n-1] | ≪ s[i]:= (a[i] ≠ b[i]) ≠ c[i] ≫ }  +
  ≪ s[n]:= d[n] ≫
END
```

Figure 4.9 Chandy-Misra adder

Layout

One problem with the fast adder presented in section 4.3.2 is the many crossing wires required to realize the transitions

$$\{ +i : [1..n] \mid \ll (i \geq t)\ AND\ (d[i] = U) \text{->}\ d[i] := d[i-t] \gg \}$$

This computation is called carry combination, and to realize it, there must be connections to $d[i]$ from $d[i\text{–}1]$, $d[i\text{–}2]$, ... and to $d[i\text{–}1]$ from $d[i\text{–}2]$, $d[i\text{–}3]$, ..., etc. Laying out these connections on an integrated circuit is both complicated and area consuming. Fortunately, it is possible to organize the computation a little differently so most of the crossing wires are avoided. To see how, take a close look at how the most significant carry $d[n]$ is determined. It is computed from $d[n/2]$, which in turn is computed from $d[n/4]$. The diagram shown below illustrates the flow of computation in the carry combination for the case $n = 8$.

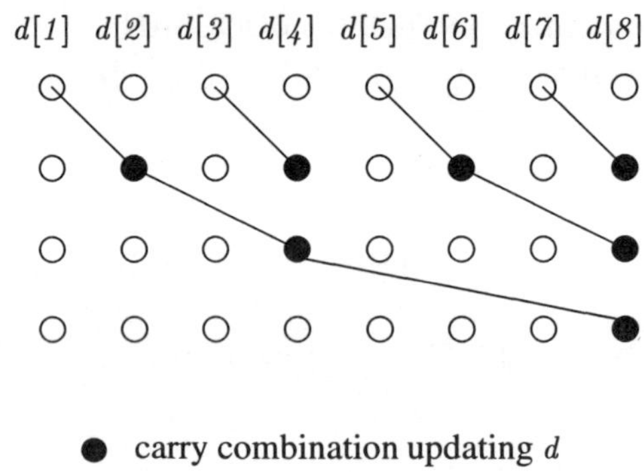

A similar analysis shows how to do the carry combination for the elements $d[1], d[2], d[4], \ldots, d[i]$. The lines in the diagram show which connections that are necessary (the solid lines). Note that no crossings are needed.

A similar analysis shows how to do the carry combination for all the remaining positions using another binary tree propagating the carry from the most significant to the least significant bits.

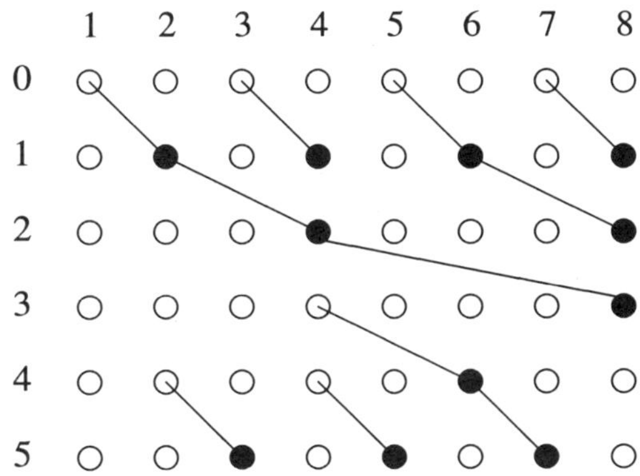

Figure 4.10 Flow of carry computations in a Brent-Kung adder

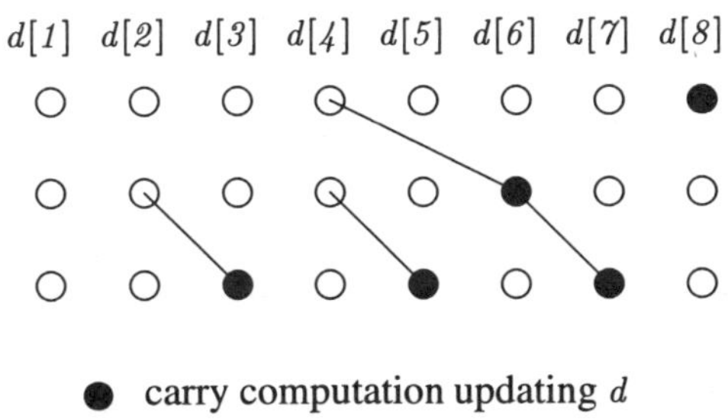

The transitions updating d perform the carry combination exactly the same way as in the Chandy-Misra adder.

$\ll d[i] = U \text{ -> } d[i] := d[i-j] \gg$

The two trees described above can be laid out by placing them next to each other in a two-dimensional structure as shown in figure 4.10. This structure is described by a two-dimensional array representing the mesh. The solid lines indicate where the carry combination is done. At all the remaining mesh points the carry value is just copied from one row to the next. Figure 4.11 shows a complete design of the carry computation. The cell *combine* does carry combination, and the cell *copy* just copies the value from one row of the mesh to the next.

4.3.4 Representing the Value U

A binary representation of d is obtained by splitting it into two vectors of booleans, called p and g, and using the following encoding:

$p[i]$	$g[i]$	$d[i]$
0	*0*	*0*
0	*1*	*1*
1	-	*U*

As is the case with d, the pair (p, g) represents the carry based on knowledge about the value of the operands in a segment of indices. Consider two such adjacent segments.

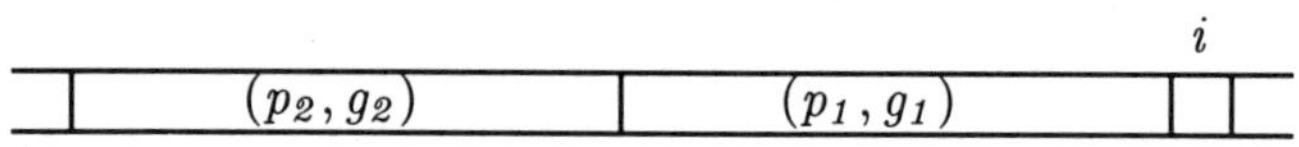

Now the pair (p, g) for the entire segment

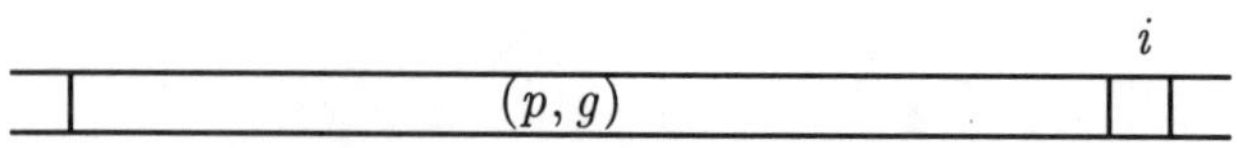

can be determined by combining (p_1, g_1) and (p_2, g_2) as follows:

$$p = p_1 \ AND \ p_2$$
$$g = g_1 \ OR \ (p_1 \ AND \ g_2)$$

This combination corresponds to the transition:

$$\ll d[i] = U \text{ -> } d[i] := d[i-t] \gg$$

This can be introduced in the design shown in figure 4.11 by changing the type of d. Instead of using a three valued type, a record type is used instead.

TYPE carryval = RECORD p, g : BOOLEAN END

The only changes that needs to be made in the design description are in the cells *combine* and *copy* which become:

```
STATIC  U = 2   logn = log(n)   h = 2*logn-2
TYPE
    wordrange = [0..n]
    wordval = [0..1]
    word = ARRAY wordrange OF wordval
    carryval = [0..2]  (* 0, 1, U *)
    carryword = ARRAY wordrange OF carryval
STATE
    a, b, s: word
    d: carryword
CELL carry(a, b: word; d: carryword)
TYPE  mesh = ARRAY [0..h+1] OF carryword
  STATE   dd: mesh
  CELL combine(c1, c2, c: carryval)
     BEGIN  ≪ c2=U -> c:= c1 ≫ END combine
  CELL copy(c1, c: carryval) BEGIN  ≪ c:= c1 ≫  END copy
BEGIN
   {+ i: [0..n-1] | ≪ dd[0,i+1]:= IF a[i] = b[i] THEN a[i] ELSE U ≫ } +
   (* first binary tree propagating values upwards *)
    {+ k: [0..logn-1] | {+ i: [1..n] |
     {i MOD 2^(k+1) = 0 | combine(dd[k,i-2^k], dd[k,i], dd[k+1,i])} +
     {i MOD 2^(k+1) ≠ 0 | copy(dd[k,i], dd[k+1,i])}
    }} +
   (* second binary tree propagating values downwards *)
    {+ k: [logn..h] | {+ i: [1..n] |
     {i MOD 2^(h-k+1)≠0 OR i>n-2^(h-k) | copy(dd[k,i], dd[k+1,i]) } +
     {i MOD 2^(h-k+1)=0 AND i≤n-2^(h-k) |
                    combine(dd[k,i],dd[k,i+2^(h-k)],dd[k+1,i+2^(h-k)])}
    }} +
    {+ i: [1..n] | ≪ d[i]:= dd[h+1,i] ≫ }
END carry
```

Figure 4.11 Carry computation of Brent-Kung adder

```
CELL combine(c1, c2, c : carryval)
BEGIN
  ≪ c.p := c1.p AND c2.p ≫ +
  ≪ c.g := c1.g OR (c1.p AND c2.g) ≫
END combine

CELL copy(c1, c : carryval)
BEGIN
  ≪ c := c1 ≫
END copy
```

4.4 BACKGROUND

Synchronous circuits are by far the most common, and therefore, most methods, tools, and hardware description languages are aimed at synchronous computations. In some cases, the nature of the problem requires a synchronous model, this is for example the case for many systolic algorithms, but in other cases, the synchronous model is used because of tradition rather than necessity. The synchronous model of parallelism is a well understood and well developed model at the circuit level. However, when used at a high-level it often means that the timing details of the low-level realization are reflected in the high-level description.

A number of systolic algorithms have been proposed and analyzed over the years. The FIR filter is often used as an introduction because of its simplicity. The design presented in this chapter is not particularly efficient and better designs have been proposed. This area of research was initiated by H.T. Kung [41], and the book [42] gives a comprehensive presentation of systolic algorithms.

The fast adder is also discussed in section 6.9 of [10] where an algorithm is presented which is similar to the one shown in figure 4.9. The Brent-Kung adder is described in [7].

4.5 EXERCISES

Exercise 4.1 Design an externally controlled shift register cell:

> *TYPE*
> $vector$ = *ARRAY* [*1..n*] *OF BOOLEAN*
> *CELL* $shift(enable$: *BOOLEAN;* v : $vector)$

The shift cell should make a cyclic shift of v whenever $enable$ is true. When $enable$ is false, v should not change.

Exercise 4.2 Design an n-bit counter, where $n > 0$ is an integer.

> *TYPE*
> $range = [0..n-1]$
> $word$ = *ARRAY* $range$ *OF BOOLEAN*
> *CELL* $counter(incr$: *BOOLEAN;* $v : word)$

When $incr$ is true the counter should increment the value in v and set $incr$ to false. The value in v is a binary representation of the value of the counter. The initial value of the counter must be 0, and when the counter reaches the value $2^n - 1$ the next increment will make it 0 again.

Exercise 4.3 Given an array a with the index range $[0..n]$, give a design that computes a matrix r with the index range $[0..n, 0..n]$, where row k of r is obtained by a circular shift of a to the right by k positions for all k, $0 \leq k \leq n$.

Exercise 4.4 Design a simple circuit for a timeout function using Synchronized Transitions. The timeout unit has three inputs $init$, in and $tick$.

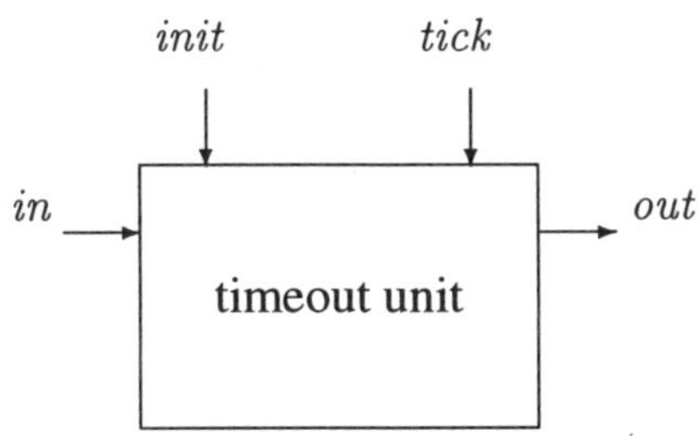

$tick$ is a periodic signal signal (could come from a clock). After initialization with $init$, the timeout unit must set out to true either when in becomes true or after *limit* periods. After this the timeout unit is inactive until the next initialization with *init*.

Exercise 4.5 Show that the cell *display* shown in figure 4.2 meets the exclusive write condition.

Exercise 4.6 Design a synchronous computation of the maximum, see section 3.1.3. Find an invariant and show that the design maintains this invariant.

Exercise 4.7 Design a revised version of the systolic filter shown in figure 4.5 where it is possible to input the filter coefficients, $a_1, a_2, \ldots, a_k$. This should be done through the input variable *xin* and controlled by an extra parameter to the cell called *load*.

Exercise 4.8 Design a vector multiplier with the following interface:

```
TYPE
  bit = [0..1]
  range = [0..d-1]
  word = ARRAY range OF bit
CELL mult(ain, win, s : word; load : BOOLEAN)
```

When *load* is true, new values of *ain*, *win* are provided. The cell *mult* should then compute $ain * win$ and deliver the result in s.

Exercise 4.9 Consider the stop watch designed in section 4.1.1. If the start/stop button is held (*ss* is true) for several consecutive clock periods, this is interpreted as if the button was pushed a number of times. Suggest a different design where holding the button for a number of periods is treated as a single push of the button.

Exercise 4.10 Show that the Brent-Kung adder maintains the following invariant (c is of type *carryval*):

INVARIANT $c.p \Rightarrow NOT\, c.g$

5

SYNCHRONOUS REALIZATIONS

The previous chapters show how to use Synchronized Transitions to model the computation of an integrated circuit and how to verify properties of the computation. This leads to a design description capturing the intended behavior. This chapter shows how to synthesize a circuit realization from such a design description. The emphasis is on realizations where there is a one-to-one correspondence between the design description and the synthesized circuits; this is called a direct realization. Each transition is synthesized separately and becomes an independent subcircuit. This means that synthesis can be done in a way similar to the one used for verification, namely by treating each transition separately. Since an independent subcircuit is synthesized for each transition, it means that the subcircuit is only active when the corresponding transition is executed. This might not be an optimal utilization of the circuitry. Sometimes a smaller (less circuitry) can be obtained by multiplexing an expensive subcircuit instead of generating a separate copy for each occurrence in the design description. Much research effort is currently devoted to such optimizations [8], and the results can be used for synthesizing designs in Synchronized Transitions; however, this is outside the scope of this chapter, where the emphasis is on direct realizations. It is possible to synthesize both synchronous and asynchronous/self-timed circuits using direct realizations. Here the emphasis is on showing how both synchronous and asynchronous designs are realized with synchronous circuitry; chapter 7 describes self-timed realizations.

It is important to note that exactly the *same* design description can be used both for verification and for synthesis. It is, therefore, not necessary to manually transform the design description in order to do automatic synthesis. This would be a likely source of errors.

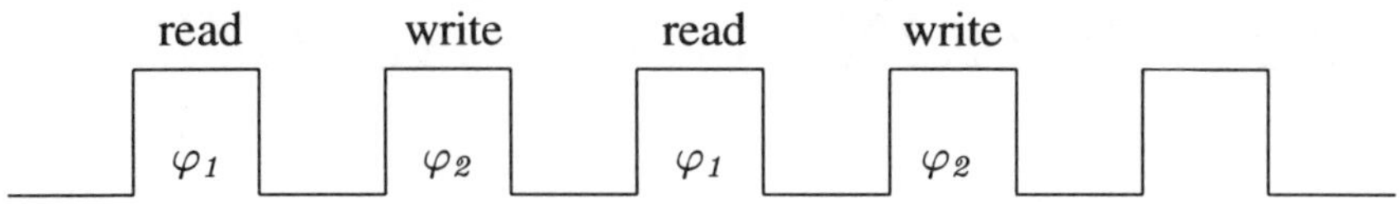

Figure 5.1 Two-phase clock

5.1 TWO-PHASE REALIZATIONS

A synchronous circuit is controlled by a global clock, which is a periodic signal. In a direct realization there is an independent subcircuit corresponding to every transition. The clock signal is distributed to all subcircuits, and it activates all of them simultaneously. Synchronous execution of transitions consists of two phases, where the first is evaluation of preconditions and right-hand-side expressions, and the second is assignment of new values to the state variables appearing on the left-hand side of active transitions. The two phases are implemented by letting every other clock cycle, called φ_1, activate the evaluations and letting the other half, called φ_2, activate the assignments (see figure 5.1). This is called a two-phase realization: The following generic transition, t, is used to illustrate the two-phase realization.

TRANSITION $t \ll c \mathbin{-\!\!>} v := e \gg$

The values of state variables are supplied on wires that are read to evaluate the precondition, c, and the right-hand side, e, in phase φ_1. Writing the variable(s) of v is done in phase φ_2. Figure 5.2 sketches a two-phase realization of t. $\boxed{c}$ is a combinational network that computes the precondition, $\boxed{e}$ is another combinational network that computes the values of the right-hand sides of the multi-assignment, and $\boxed{v\!:=}$ stores the new values of the state variables; further

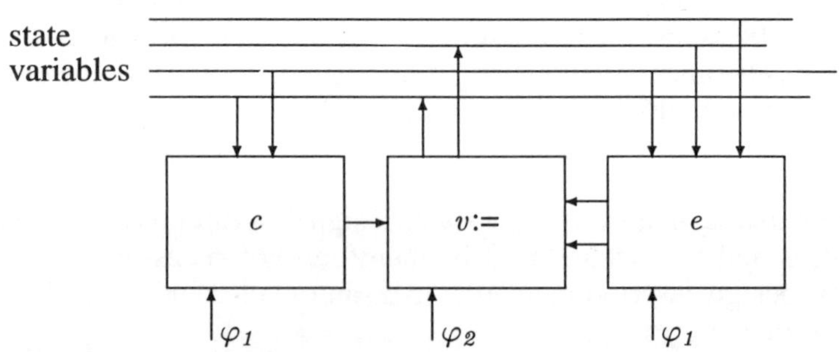

Figure 5.2 Two-phase realization of a single transition

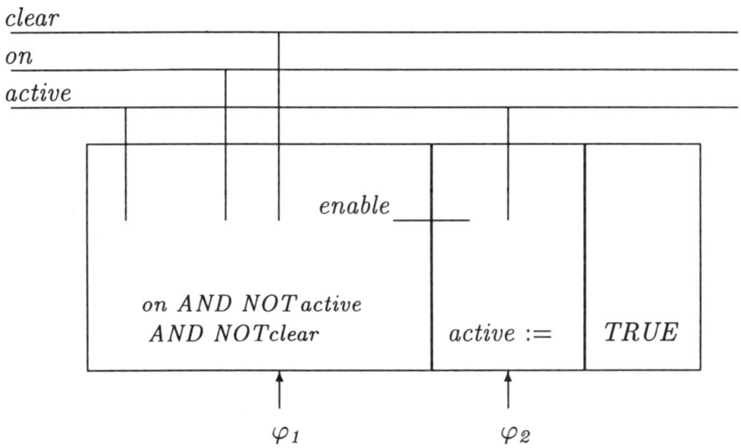

Figure 5.3 Two-phase realization of a transition

details are given below. The three components of a transition, $\boxed{c}$, $\boxed{e}$, and $\boxed{v{:=}}$, always follow the same pattern, but the internal function is different for different transitions. Figure 5.3 shows the two-phase realization of the following transition:

$$\ll on\ AND\ NOT active\ AND\ NOT clear \text{ -> } active := TRUE \gg$$

A standard logic synthesis program [6] can be used to generate the combinational logic needed to realize the precondition and the right-hand side. The details are not discussed here.

5.1.1 State Variables

In figure 5.2 the values of state variables are represented on wires. These wires connect all subcircuits reading or writing a particular state variable. In addition to the wire, a storage mechanism is required to hold the current value of the variable. There are a number of ways to realize such a storage mechanism; however, the following two characteristics are important.

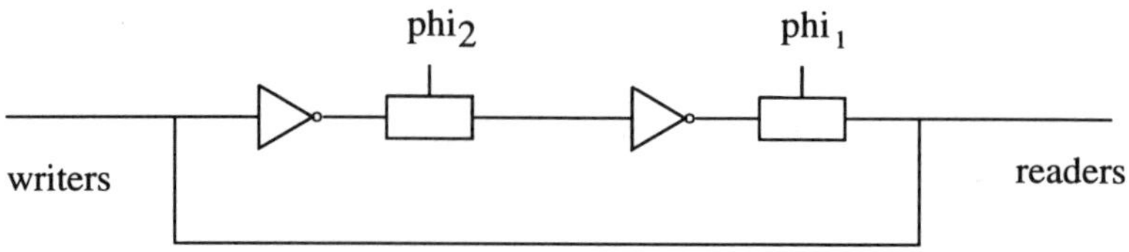

Figure 5.4 Dynamic refresh element

- The values of state variables are changed by many different transitions.
- No upper bound can be given on how long the values of state variables must be retained.

The transitions are synthesized as independent subcircuits; when realizing one particular transition, no knowledge is required of other transitions. This precludes realizations where a state variable is associated with a single transition. Instead, the value of a state variable is retained on a wire (in case of a boolean variable) or on a number of wires (other types). The capacitance of the wire can hold the value written to the wire for a short period of time; however, some refresh circuitry is needed to keep the value stable when it is not written. In full-custom VLSI circuits, a refresh element can be realized by a "bleeder" where the transistors are sized appropriately. Alternatively, a clocked refresh element can be used. A schematic diagram is shown in figure 5.4. The refresh element reassigns the current value of the state variable in phase φ_1, while a new value can be written in phase φ_2. The refresh element consists of two cascade coupled clocked inverters. This realization has been implemented and analyzed by Jacob Christensen; further details are given in [11], which describes a prototype tool that automatically generates a VLSI layout (using the two-phase realization) from a synchronous design described in Synchronized Transitions.

5.1.2 Combining Transition Realizations

A synchronous design with many transitions is realized by synthesizing a subcircuit like the one shown in figure 5.2 for each transition. Each subcircuit reads some of the state variables and writes new values to one or more state variables. The complete design is realized by connecting all these subcircuits, i.e., connecting all the wires from all subcircuits corresponding to a particular state variable. This is a standard routing problem that can be solved with

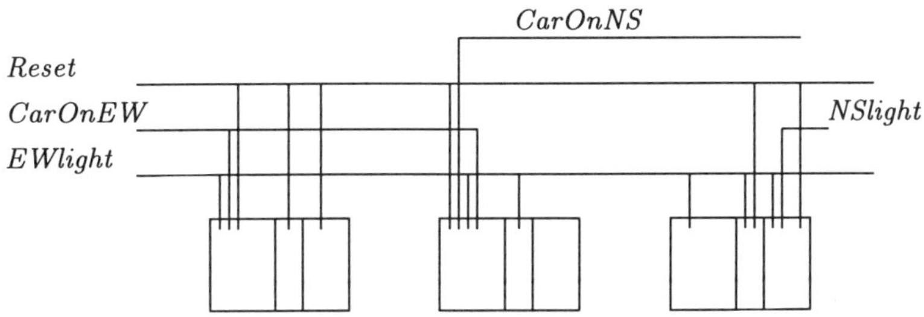

Figure 5.5 Direct connection of two-phase realizations

a channel router; an example is shown in figure 5.5. The cell mechanism encourages locality, which might improve layouts, because local variables are only needed in a small part of the design.

In a synchronous circuit, the clock frequency must be low enough to allow all subcircuits to finish their computation within a clock period. When a synchronous design is realized by simply connecting a number of two-phase realizations, the preconditions and right-hand sides of all transitions are evaluated in parallel. Therefore, the clock period must be long enough to allow the slowest of these to finish; in section 5.2 it is shown how to estimate the clock period of a two-phase realization.

5.1.3 Evaluation

The two-phase realization is very simple compared to the sophisticated techniques used in state-of-the-art synthesis tools. In many cases a direct realization is not efficient, because a separate copy of the necessary hardware is generated for each transition or cell. In state-of-the-art high-level synthesis tools [8], advanced allocation and scheduling methods are capable of multiplexing the same hardware units for many different purposes. These methods could also be used for synthesizing efficient circuits from design descriptions in Synchronized Transitions, although no tools are currently available for doing this.

However, the direct correspondence between a design description and the realization also has some advantages. In the next section on timing estimation, it is illustrated how one can take advantage of this correspondence and use the design description for identifying bottlenecks and estimating the clock period of a design.

5.2 TIMING ESTIMATION

In this section, it is shown how to estimate the clock period from a design description. This enables the designer to evaluate the performance of a design at an early stage in its development. Experience with the technique indicates that there is a good correspondence between the estimates and the actual execution time of the synthesized circuit [11]. The estimation technique is based on the one-to-one correspondence between the design description and the realization, which is obtained by using a direct realization, such as the two-phase realization introduced in section 5.1.

5.2.1 Critical Paths

In a synchronous circuit, the computations of all subcircuits must be finished within one clock period. Therefore, the length of the clock period is determined by the slowest subcircuit, called the critical path. Locating the critical path allows one to adjust or redesign the speed-limiting parts and thereby improve overall performance.

Direct realizations make it possible to identify critical paths by analyzing the design description. Consider the following transition, t :

$$\ll c -> v := e \gg$$

The computations of t are done by the combinational logic realizing the precondition c and the right-hand side e. These are evaluated in parallel. The complexity of the logic computing c and e determines the time it takes to execute the transition. Therefore, the critical path can be identified by analyzing the design description and pointing out the transition with the most time-consuming precondition or right-hand side. Section 5.2.2 discusses how this can be estimated. The estimation technique assumes that gate-delays dominate wire-delays.

5.2.2 Approximation of the Clock Period

The clock period of a direct realization depends on the computation time of the slowest transition. This computation time is determined by the complexity of the expressions appearing in preconditions and right-hand sides of assignments, which can be estimated from their syntactical structure as explained below for boolean expressions. Consider a boolean expression, E, and let $D(E)$ denote the estimate of the computation time of E. The syntactical structure of E is used to derive $D(E)$. Below, four syntactical cases are distinguished, and $\oplus$ denotes an arbitrary dyadic operator, e.g., AND ; $\ominus$ denotes a monadic operator, e.g., NOT:

- E is of the form $x \oplus y$, where x and y are themselves boolean expressions. The estimate $D(E)$ is $max(D(x), D(y)) + T_{\oplus}$, i.e., the maximum of the computation times of x and y plus the time to evaluate $\oplus$.
- E is of the form $\ominus x$, where x is a boolean expression. The estimate $D(E)$ is $D(x) + T_{\ominus}$.
- E is the name of a state variable. The estimate $D(E)$ is a constant T_r (time to read a state variable).
- E is is a constant ($TRUE$ or $FALSE$). The estimate $D(E)$ is a constant T_c (time to generate a constant).

The absolute values of the quantities $T_{\oplus}$, $T_{\ominus}$, T_r, and T_c are technology dependent. It is not necessary to know the exact values to identify the critical paths of a design and to balance it. In [11] the values are determined experimentally and used to analyze a number of designs. Consider the following expression, E:

$$a\ AND\ (b\ OR\ NOT(c\ AND\ d))$$

E has four levels of logic, as shown by the expression tree in figure 5.6. This tree reflects the physical realization. For simplicity, $T_{\oplus}$ and $T_{\ominus}$ are assumed to be 1 time unit, and T_r and T_c are assumed to be 0. For E the computation time is

$$\begin{aligned} D(E) &= max(D(a), D((b\ OR\ NOT(c\ AND\ d)))) + T_{AND} \\ &= D((b\ OR\ NOT(c\ AND\ d))) + T_{AND} \\ &= max(D(b), D(NOT(c\ AND\ d))) + T_{OR} + T_{AND} \\ &\dots \end{aligned}$$

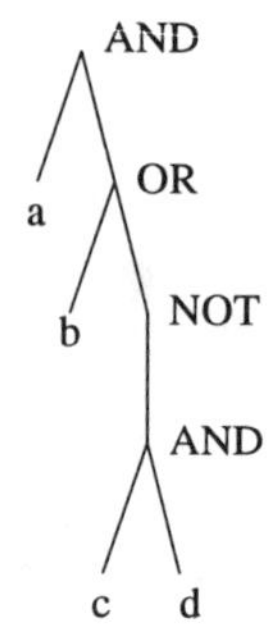

Figure 5.6 Boolean expression tree

$$
\begin{aligned}
&= T_{AND} + T_{NOT} + T_{OR} + T_{AND} \\
&= 4
\end{aligned}
$$

The performance observations acquired at the source level can be used to improve the designs, for example, by identifying bottlenecks caused by preconditions or right-hand sides that are much slower than expressions found in the rest of a design. In [11] this approach is generalized to integer expressions. However, more implementation-dependent calibration is needed to get useful estimates of bottlenecks and computation times of integer expressions.

In this section, it has been demonstrated that it is possible to estimate the performance of a design based on a high-level description in Synchronized Transitions. The prerequisite is, however, that a direct realization is used. If the synthesis tools perform major reorganizations of the design in order to optimize the final circuit, the high-level description is no longer an adequate model of the computation time, and the estimation technique cannot be used.

5.2.3 Balanced Designs

The best utilization of the synthesized circuitry is obtained when it takes the same time to execute all transitions of a design, because then all circuitry is constantly active. For such designs, a direct realization evaluates all preconditions and right-hand sides in parallel, and no subcircuits are inactive waiting for others to finish. This gives a perfect utilization of the hardware. This ideal is called a balanced design, and in such a design all the preconditions and right-hand sides in the design take the same time to evaluate. Complete balance is rarely achieved in practical designs, but it is a useful goal for guiding the designer.

Consider the two designs t_1+t_2 and t_2+t_3 using the following three transitions:

TRANSITION t_1 $\ll x$ OR y $->$ $z := NOTx \gg$
TRANSITION t_2 $\ll x$ AND y $->$ $z := z$ OR $y \gg$
TRANSITION t_3 $\ll (x$ AND $y)$ OR $(z$ AND $y)$ $->$ $c := TRUE \gg$

The four expressions appearing in t_1 and t_2 take roughly the same time to execute, so t_1+t_2 is a balanced design. Considering t_3, the precondition is more complex than the one found in t_2, while the right-hand side is simpler. Therefore, $t_2 + t_3$ is unbalanced.

Most practical designs have some imbalance, but as a rule of thumb, it should be minimized by shortening time consuming transitions. If a design is unbalanced, transitions on the critical path can be divided. To illustrate this, consider the transition, t_{skew}:

TRANSITION t_{skew} $\ll x + y \geq z$ $->$ $a := FALSE \gg$

The precondition of this transition contains the expression $x + y \geq z$, which requires both an addition and a comparison. The same result is obtained with the following two transitions:

TRANSITION $t_1 \ll v := x + y \gg$
TRANSITION $t_2 \ll v \geq z$ $->$ $a := FALSE \gg$

The transition t_{skew} performs the same function as $t_1 + t_2$; however, the final result appears in a one iteration later. This type of transformation, often called pipelining, is very common, and is useful for improving the balance of a design. Note that pipelining is done in the high-level design description, and hence, it is possible to verify it or experiment with the effects before attempting a low-level circuit realization. Pipelining an unbalanced transition is not always desirable. It is a tradeoff between area, speed, and readability. In the case of t_{skew} an extra state variable v is introduced, and the result is delayed one clock period.

5.3 ASYNCHRONOUS DESIGNS

Many designs are simple to describe and verify as asynchronous computations. This is illustrated by several examples given in the previous chapters and in [10]. As explained in chapter 7, such asynchronous designs can be implemented with

```
CELL example
  STATE a, b: BOOLEAN
  INVARIANT  a OR b
BEGIN
  ≪ a:= NOT b ≫  ||  ≪ b:= NOT a ≫
END example
```

Figure 5.7 A design that cannot be directly realized

asynchronous circuitry, but synchronous circuits are by far the most common which makes it important to be able to implement asynchronous designs by synchronous circuitry. This section describes how this can be done using the two-phase realization introduced in section 5.1.

In general, it is, however, not possible to use a two-phase realization and preserve the atomic behavior of transitions in an asynchronous design. In the two-phase realization, all enabled transitions are executed in parallel in each clock period. In many cases, such a simultaneous execution of transitions is not equivalent to the atomic execution (one transition at a time), which is the abstract model of an asynchronous computation. This is illustrated by the design shown in figure 5.7. It is simple to verify that this design maintains the invariant. Hence, atomic execution never leads to states where it is violated. However, simultaneous execution of the two transitions may lead to states that do not satisfy the invariant. For example, in a state where the two state variables a and b are true, simultaneous execution of both transitions results in a state where both a and b are false, violating the invariant. In a two-phase realization, transitions are executed simultaneously, and hence, this is not always a correct realization. There is a fundamental discrepancy between the abstract model of asynchronous designs where transitions are atomic and the two-phase realization where they are executed simultaneously. Although the two-phase realization cannot be used for all designs, e.g., the one shown in figure 5.7, there is a significant class of asynchronous designs for which it is correct. This class is characterized in the next section.

5.4 IMPLEMENTATION CONDITIONS

This section shows how to avoid the fundamental discrepancy between the abstract model of asynchronous designs, where the transitions are atomic, and realizations where they are performed simultaneously. One could use a realization that performs the transitions one at a time, as prescribed by the abstract model. But this inevitably requires some additional circuitry to ensure the atomicity and yields less efficient circuits. Instead, the problem is turned upside down by asking: Is there an interesting class of designs for which the two-phase realization is usable? It turns out that the answer is yes. So instead of changing the realization, a restricted class of designs is defined for which the two-phase realization is correct. The restricted class is characterized by a condition that must be met by the design description. Such a condition is called an implementation condition. A similar approach based on restricting the design is used for other realizations. There are, for example, implementation conditions for self-timed realizations; these are described in chapter 7.

5.4.1 The Implementation Condition CREW

This section presents an implementation condition, called Concurrent Read Exclusive Write or just CREW, which is sufficient to ensure that the two-phase realization is a correct implementation of an asynchronous design. The condition can be checked mechanically in a way very similar to the one used for verifying invariants and protocols. Intuitively, the CREW condition ensures that even if transitions are executed simultaneously, as is the case in a two-phase realization, then no states are entered which cannot also be entered by execution the transitions one at a time.

In a two-phase realization, *all* enabled transitions are executed in *every* clock period. As illustrated by the example shown in figure 5.7, this is not always a correct way of realizing an asynchronous design. In the example, the problem is that each of the two transitions reads the value of a state variable that is written by the other transition. The implementation condition excludes designs with such transitions. To define the condition formally, consider a transition, t:

$$\textit{TRANSITION}\ t \quad \ll c \rightarrow v_1, v_2, \ldots, v_k := e_1, e_2, \ldots, e_k \gg$$

The write set of t, W^t, is $\{v_1, v_2, \ldots, v_k\}$. Similarly, the read set, R^t, is the set of variables appearing in the expressions c and $e_1, e_2, \ldots, e_k$. The predicate $active(t)$ is a shorthand for the assertion

$$c \wedge ((v_1 \neq e_1) \vee (v_2 \neq e_2) \vee \ldots (v_k \neq e_k))$$

The definitions given below all refer to an asynchronous design D:

$$D: \quad t_1 \ || \ t_2 \ || \ \ldots \ || \ t_N$$

Definition 5.1 *The read/write graph of D is a directed graph with one node for each transition t_i and edges E, where E is given by*

$$(t_i, t_j) \in E \quad \textit{if and only if} \quad W^{t_i} \cap R^{t_j} \neq \emptyset \wedge i \neq j$$

That is, there is an edge from t_i to t_j if t_i writes into a variable read by t_j, i.e.,if executing t_i may influence t_j. Note that the definition requires $i \neq j$; hence the read/write graph does not contain any edges starting and ending at the same node (corresponding to transitions such as $\ll t := t + 1 \gg$).

Definition 5.2 *A read/write cycle in D is a nonempty set of transitions whose corresponding nodes form a cycle in the read/write graph.*

Definition 5.3 *A minimal read/write cycle in D is a read/write cycle that does not contain another read/write cycle as a proper subset.*

The existence of a read/write cycle in the read/write graph indicates a potential conflict when trying to execute transitions simultaneously. However, if there are one or more of the transitions on the cycle that are not active, then the conflict is harmless. This intuition is expressed more formally in the implementation condition CREW (Concurrent Read, Exclusive Write):

Definition 5.4 *A design, D, meets the implementation condition CREW, if and only if the following assertion is satisfied for all minimal read/write cycles $rwc = \{t_1, t_2, \ldots, t_k\}$ of D:*

$$CREW(rwc) \equiv \neg(active(t_1) \wedge active(t_2) \ldots active(t_k))$$

Establishing $CREW(rwc)$ may require more analysis than just a local argument using boolean algebra. Typically, the invariant of the design is needed.

Example: The Traffic Light Controller

This section illustrates how to check the CREW condition by showing that the simple traffic light controller meets this implementation condition. The four transitions of the traffic light controller are

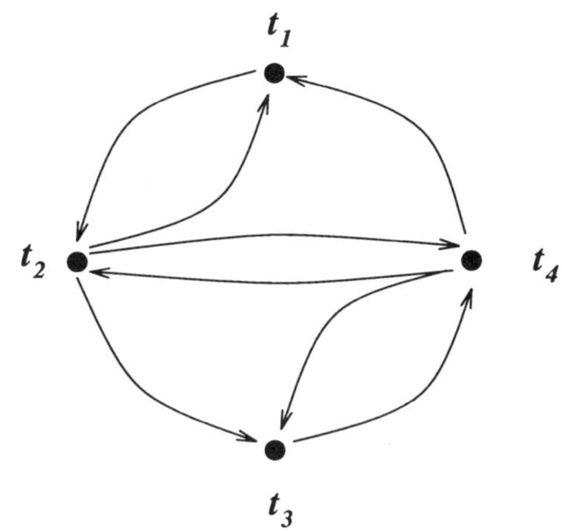

Figure 5.8 Read/write graph for the traffic light controller

t_1: $\ll$ (*EWlight* = *green*) *AND CarOnNS* –> *EWlight* := *yellow* $\gg$ ||
t_2: $\ll$ (*EWlight* = *yellow*) –> *EWlight*, *NSlight* := *red*, *green* $\gg$ ||
t_3: $\ll$ (*NSlight* = *green*) *AND CarOnEW* –> *NSlight* := *yellow* $\gg$ ||
t_4: $\ll$ (*NSlight* = *yellow*) –> *NSlight*, *EWlight* := *red*, *green* $\gg$

The read/write graph shown in figure 5.8 has four nodes, one for each of the four transitions. There are three minimal read/write cycles: $t_1 t_2$, $t_2 t_4$, and $t_3 t_4$. Note that there are many other cycles in the graph, for example $t_1 t_2 t_4$, but these are not minimal cycles. The CREW condition corresponding to the three minimal cycles are:

1) ¬(((*EWlight* = *green*) *AND CarOnNS AND* (*EWlight*≠*yellow*)) *AND*
((*EWlight* = *yellow*) *AND* (*EWlight*≠*red*) *AND* (*NSlight*≠*green*)))

2) ¬(((*EWlight* = *yellow*) *AND* (*EWlight*≠*red*) *AND* (*NSlight*≠*green*)) *AND*
((*NSlight* = *yellow*) *AND* (*NSlight*≠*red*) *AND* (*EWlight*≠*green*)))

3) ¬(((*NSlight* = *green*) *AND CarOnEW AND* (*NSlight*≠*yellow*)) *AND*
((*NSlight* = *yellow*) *AND* (*NSlight*≠*red*) *AND* (*EWlight*≠*green*)))

It is very simple to see that the first and third assertions hold, because a state variable cannot have two different values. The second condition could be violated if *NSlight* and *EWlight* both have the value *yellow*. However, from the invariant it follows that either *NSlight* or *EWlight* are red. Hence, the second assertion also holds. Therefore, the traffic light design meets the CREW condition.

5.4.2 Soundness of CREW

CREW ensures that the states that can be reached by a synchronous computation can also be reached by an atomic execution of the same transitions. To see that this is the case, consider one step of a synchronous computation starting in a state σ and leading to a state σ'.

$$\sigma \qquad\qquad\qquad\qquad \sigma'$$

$$t_1 : \ll \; \dots \; \gg$$

$$t_2 : \ll \dots \gg$$

$$t_n : \ll \; \dots \; \gg$$

Below it is shown that there is a permutation of the transitions enabled in σ whose atomic execution also leads to σ'. Hence, any state that can be reached by the synchronous computation can also be reached by the atomic computation.

Consider the read/write graph. From the CREW condition it follows that:

$$\neg(active(t_1) \wedge active(t_2) \dots active(t_k))$$

i.e., that in state σ, in any minimal read/write cycle there is at least one transition t where either

- t is not enabled in σ, or
- execution of t in σ leaves the state unchanged.

For any such t, removing the outgoing edges from t in the read/write graph reflects that in the actual state σ it cannot influence other transitions. This breaks all minimal cycles. Since *all* outgoing edges from these transitions are removed, also nonminimal cycles are broken. Therefore, a resulting acyclic graph is obtained where the absence of a path from t_i to t_j means that the execution of t_i in σ does not affect the enabledness or effect (change of variables) of t_j. Since the graph is acyclic, there exists a linear ordering $t_{i_1} \prec t_{i_2} \prec \dots \prec t_{i_N}$ of all the transitions such that $t_i \prec t_j$ implies that there is no directed path from t_i to t_j and thus t_i cannot affect t_j. It follows that a sequential execution of $t_{i_1}, t_{i_2}, \dots, t_{i_N}$ (skipping transitions not enabled) starting from σ has the same effect as their simultaneous execution. Therefore, any state that is obtained by a two-phase realization of a design meeting CREW can also be obtained by executing the transitions one at a time (atomically). Therefore, it is also ensured that any invariant of the original design is maintained by the two-phase realization.

It should be noted that CREW is only a sufficient condition for guaranteeing the correctness of the two-phase realization; it is not necessary. Therefore one can find designs that do not meet the CREW condition for which the two-phase realization is nevertheless correct. From a theoretical point of view, this might seem unsatisfactory; however, for practical usefulness, it is more important to have a condition that is easy to check than to cover all possible cases, many of which may not occur in practical designs. CREW is formulated in such a way that it is simple to check it mechanically. This is discussed further in the next section.

5.4.3 Checking CREW Mechanically

To check that a large design meets the CREW condition, it is necessary to verify a large number of instances of the formula $\mathit{CREW}(\mathit{rwc})$. This is usually quite tedious and therefore error prone. Furthermore, the conditions have to be rechecked every time a small change is made to the design. However, it is possible to mechanize the check in the same way that the verification of invariants and protocols is mechanized. The key step of the CREW check is to prove an assertion of the form

$$\neg(\mathit{active}(t_1) \wedge \mathit{active}(t_2) \ldots \mathit{active}(t_k))$$

One such assertion must be checked for each minimal read/write cycle of the design. To automate the check of CREW, the first step is therefore to find minimal read/write cycles. First, the read/write graph is generated from the design description in Synchronized Transitions. From this graph the minimal cycles are computed. There are several algorithms for doing this (see [37]). In general, the number of cycles in a graph grows exponentially. However, this is usually not a serious problem for practical designs that exhibit a lot of locality. The generation of the minimal cycles is done by a preprocessor. Once they are computed, the condition $\mathit{CREW}(\mathit{rwc})$ is generated for each minimal cycle rwc. Structurally, this is a simple formula; to prove it, it is usually necessary to exploit additional properties of the design, for example, declarations and invariants. Therefore, the CREW condition is augmented with other information generated by the translator ST2LP before generating proof

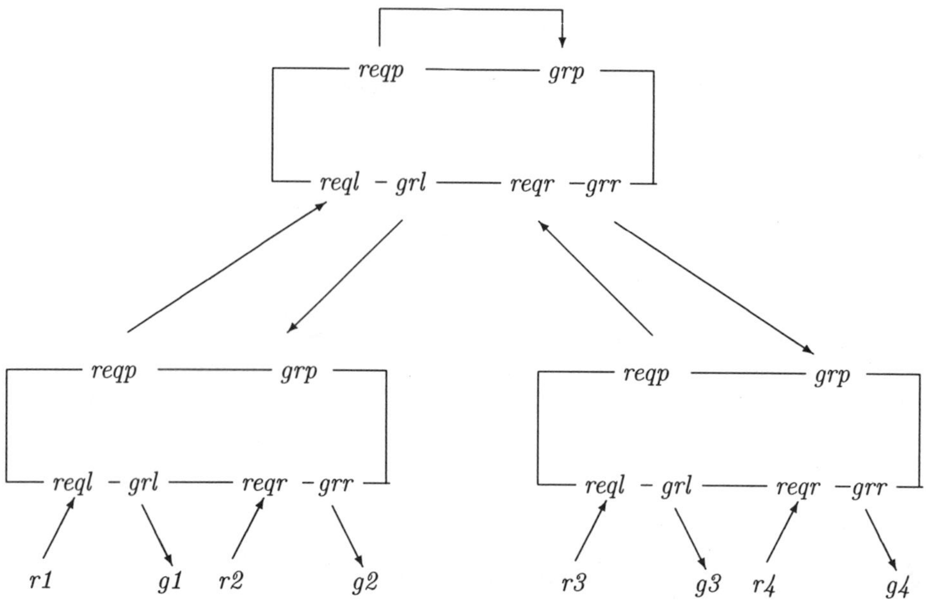

Figure 5.9 Two levels of arbiter tree

obligations for the theorem prover.

Example: The Arbiter

As shown below, the arbiter introduced in section 2.9 meets the CREW condition, and hence, it can be implemented with a two-phase realization. Figure 5.9 shows an arbiter with three nodes. A transition at the root copies *reqp* to *grp* so that the root is able to grant its own request immediately. In this example, there are 14 minimal read/write cycles. There is, for example, a minimal cycle involving *reql* (of the parent), which is read by the *grantleft* transition where *grl* is written. The state variable *grl* is written by the *requestparent* transition of the left child, which writes *reql*. Because of this cycle, called c_1, the following condition, $CREW(c_1)$ must be checked:

$$NOT(((((grp\ AND\ reqp)\ AND\ reql)\ AND\ (NOT(grr))))\ AND$$
$$NOT(grl)\ AND\ (((NOT(grl))\ AND\ (r1\ OR\ r2)))\ AND\ NOT(reql))$$

Similar conditions must be checked for the other 13 cycles. LP has been used to successfully check each of the 14 conditions. Hence, it is safe to realize the arbiter as a synchronous circuit using the two-phase realization.

5.4.4 Making a Design Meet CREW

The CREW condition is rather strong, and many designs do not immediately meet it when they are first written down. However, there are several transformations that can be used to obtain a design meeting CREW. Consider the following two transitions:

$$\textit{TRANSITION}\ t_1 \quad \ll c_1 \ \text{->}\ v_1 := e_1 \gg$$
$$\textit{TRANSITION}\ t_2 \quad \ll c_2 \ \text{->}\ v_2 := e_2 \gg$$

Assume that t_1 and t_2 form a read/write cycle, and that the conjunction $c_1 \wedge c_2$ is satisfiable. Therefore, the two transitions do not meet CREW. To prevent t_1 and t_2 from being executed simultaneously, one can modify c_1 and c_2 and ensure that their conjunction is always false. The designer must decide which transition to execute when both preconditions are satisfied. Let the boolean expression, B, represent this decision. If B is true, the first transition is executed, and otherwise the second. To include B, the two transition are rewritten as follows:

$$\textit{TRANSITION}\ u_1 \quad \ll B\ OR\ NOT c_2 \gg * t_1$$
$$\textit{TRANSITION}\ u_2 \quad \ll NOT B\ OR\ NOT c_1 \gg * t_2$$

The preconditions of u_1 and u_2 now exclude each other, and no cycle involving these two transitions violates CREW.

There are several ways of implementing B. The simplest is to make a static choice, for example, to always execute t_1 when both are enabled. This leads to the simplified versions of u_1 and u_2 shown below:

$$\textit{TRANSITION}\ u_1 \quad t_1$$
$$\textit{TRANSITION}\ u_2 \quad \ll NOT c_1 \gg * t_2$$

If such a static choice is not adequate, one can implement a dynamic decision by letting B be a boolean state variable b assigned in the two transitions as follows:

$$\textit{TRANSITION}\ u_1 \quad \ll b\ OR\ NOT c_2 \ \text{->}\ b := FALSE \gg * t_1$$
$$\textit{TRANSITION}\ u_2 \quad \ll NOT b\ OR\ NOT c_1 \ \text{->}\ b := TRUE \gg * t_2$$

Many other variations are possible. The example presented in the next section shows how an additional state variable can be used.

Example: Sorting

Consider the design shown in figure 5.10. It describes a simple sorter using the bubble sort algorithm. At the fixed point, the elements of the vector y

```
TYPE  vector = ARRAY [1..n] OF INTEGER
STATE y: vector
FIXPOINT FORALL i: [1..n-1]. y[i] <= y[i+1]
BEGIN
  {|| i: [1..n-1] | ≪ y[i], y[i+1]:= min(y[i], y[i+1]), max(y[i], y[i+1]) ≫ }
END
```

Figure 5.10 A simple sorter

```
TYPE  vector = ARRAY [1..n] OF INTEGER
STATE
  y: vector
  turn: boolean
BEGIN
  ≪ turn:= NOT turn ≫ ||
  {|| i: [1..n-1] | ≪ (i MOD 2) = turn –> y[i], y[i+1]:=
                               min(y[i], y[i+1]), max(y[i], y[i+1]) ≫ }
END
```

Figure 5.11 A sorter that meets CREW

are sorted in increasing order. However, the design does not meet the CREW condition, because every pair of neighboring transitions form a minimal cycle in the read/write graph where the preconditions do not exclude each other.

The design can be transformed using the last transformation mentioned in section 5.4.4. An additional boolean state variable *turn* is introduced, and the transitions are divided into two groups, called odd and even transitions (determined by the range variable i in the quantified instantiation). The resulting design is shown in figure 5.11. In this design, neighboring transitions are never enabled simultaneously; when *turn* is true all even transitions are enabled, and when *turn* is false the odd transitions are enabled. Hence, the design meets CREW. The synchronous realization of this design is called odd-even transposition sort [40].

5.5 BACKGROUND

Automated synthesis is a very active area of research. There is a clear trend towards higher and higher levels of synthesis. Low-level synthesis, e.g., layout and logic synthesis, is now common practice, and it is available in many commercial products. High-level synthesis is an active area of research where impressive results have been obtained [8]. The direct realization technique described in this chapter is very simple compared to the sophisticated methods currently used for high-level synthesis. As mentioned in section 5.1.3, these methods could also be used for synthesizing design descriptions in Synchronized Transitions.

In his masters thesis [11], Jacob Christensen made a thorough investigation of the direct realization and described a prototype tool that is capable of synthesizing design description. The technique for estimating the clock period described in section 5.2 was first presented in this thesis.

The granularity of transitions is an important issue. There is a wide range of possibilities from the very fine grained where transitions describe operations on individual bits to the coarse grained where the entire computation, such as a matrix product, is described with a single transition. There are several results indicating when and how a single coarse-grained transition can be replaced with many fine-grained transitions [45]. In this chapter the reverse issue is discussed, namely, realizing the behavior of a collection of fine-grained transitions (an asynchronous design) with a single coarse-grained transition (a synchronous realization). The definition of CREW stems from [64].

5.6 EXERCISES

Exercise 5.1 Pick a technology where you have access to the performance parameters determining gate delays. Determine the constants $T_{\oplus}$, $T_{\ominus}$, T_r, and T_c that were introduced in section 5.2.2.

Exercise 5.2 Is the traffic light design presented in figure 2.1 balanced?

Exercise 5.3 Give all the cycles of the graph shown below. Indicate which of

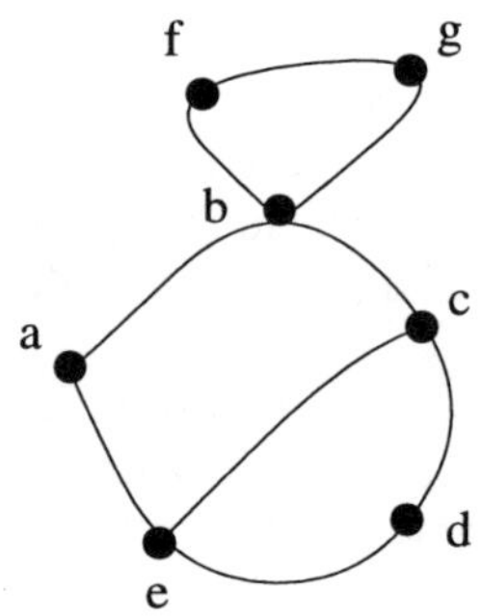

the cycles are minimal.

Exercise 5.4 Verify that the design given in figure 5.7 maintains its invariant.

Exercise 5.5 Verify that the design given in figure 5.11 meets CREW, by deriving the read/write graph and proving that this meets the CREW condition.

Exercise 5.6 Give the read/write graph for the traffic light design shown in figure 2.1. Does it meet the CREW condition?

Exercise 5.7 Suggest a design that demonstrates that CREW is not a necessary condition for the correctness of the two-phase realization, i.e., the design should violate CREW, but have a correct two-phase realization.

Exercise 5.8 Write a complete design description of a three-node arbiter tree. Give the corresponding read/write graph. List all the minimal read/write cycles. Check that the design meets the CREW condition.

Exercise 5.9 The example in section 3.1 is a design for computing the maximal element of a vector. This design does not meet CREW. Why not? Suggest a different design for computing the maximum that meets CREW.

Exercise 5.10 Verify that the ripple-carry adder shown in figure 2.4 meets CREW.

6

REFINEMENT

Abstraction mechanisms are an essential part of a high-level design technique; they enable designers to handle large and complicated problems with limited resources (both mental and physical). Abstraction does not make the complexity of large designs vanish, but it allows one to confront it in small manageable portions. When designing complicated computations, such as large integrated circuits, it is often necessary to initially abstract from efficiency considerations and concentrate on the functional correctness. Not because efficiency is unimportant, but experience has shown, that mixing efficiency and functional correctness early in the design often leads to compromising both. The carry computation shown in figure 6.1 illustrates this. On the one hand, it is an abstraction that makes it relatively easy to understand and verify the functional correctness of the Brent-Kung adder (see section 4.3.3). On the other hand the design shown in figure 6.1 is not directly suited for synthesizing an efficient circuit realization.

In general, one first makes an abstract design that describes the computation without specifying exactly *how* to realize it. When the abstract design is complete, and it has been shown that it has the desired properties, one can concentrate on finding a similar but more efficient design. This chapter presents a formal definition of what it means for one design to be "similar" to another together with verification techniques which show that designs meet the definition. The formal notion of similarity is called refinement. It is a relation between two designs called the abstract design and the concrete design. In the case of the carry computation, the adder shown in figure 6.1 is the abstract design, and the Brent-Kung adder presented in section 4.3.3 is an example of a concrete design. Often, it is not possible to go directly from the abstract design to the final concrete (circuit) design. Instead, one goes through a series

```
CELL carry(a, b: word; d: carryword)
  STATE t: INTEGER
  INITIALLY t = 1
  FUNCTION prop(a, b: word; i,t: INTEGER): BOOLEAN
    RETURN {AND j: wordrange | (i-t≤j)  AND  (j<i) => a[j] ≠ b[j] }
  INVARIANT
    FORALL i: wordrange . (0 < t) AND
      (t > i  =>  d[i] = c[i]) AND
      (t≤i  =>  d[i] = IF  prop(a, b, i, t) THEN U ELSE c[i])
BEGIN
  {+ i: [1..n] | ≪ (i≥t)  AND  (d[i]=U)  ->  d[i]:= d[i-t] ≫ }  +
  ≪ t:= 2*t ≫
END carry
```

Figure 6.1 Carry computation

of intermediate designs where each new (lower) design is more concrete than the previous one. In practise, abstract and concrete are therefore not absolute terms, what is concrete at one level is abstract at another. The more complicated a design is, the more abstraction levels are needed. To avoid unnecessary confusion, this chapter focuses on a single refinement with just two levels.

Definition of Refinement

Typically, an abstract design leaves a range of possible realizations open. There are, for example, many different ways of representing the three values used in the carry computation. In the concrete design one of the possible realizations is chosen. Therefore, a concrete design may not exhibit all the behaviors of the abstract design. This relationship is called a refinement.

> **Definition 6.1** *A concrete design is a refinement of an abstract design, if and only if, the set of computations of the concrete design is a subset of the computations of the abstract design.*

Assume that the abstract design works in a certain environment. If the abstract design is replaced by one of its refinements, then the environment will only see behaviors that it could also have seen from the abstract. It is therefore safe to substitute an abstract design by a concrete design (provided that it is a refinement). This substitution property is an important aspect of refinement. To make substitution feasible the concrete and the abstract design must be syntactically similar; if both are cells, the number and type of the formal parameters must be exactly the same (see also section 6.4). There is no requirement that the concrete design can exhibit *all* the behaviors of the abstract design. Since the emphasis in the approach described in this book is on safety, the definition of refinement gives an adequate characterization of what it means for one design to be a realization of another (see also section 6.5).

The definition does not directly suggest a useful way of verifying refinement, because it requires that all computations of both the abstract and concrete designs are described in a form that makes it possible to show that one is a subset of the other. The following sections present a more practical verification technique for showing refinement; as is the case with the other verification techniques described in this book, the technique presented here is a compromise between power and practical feasibility. The technique is not complete, which means that there are examples of designs where one is a refinement of the other, which cannot be verified with the technique described here. On the other hand, it is possible to mechanize the verification technique which is important for its practical use.

6.1 ABSTRACTION FUNCTIONS

This section introduces abstraction functions that are used for verifying that one design, the concrete, is a refinement of another, the abstract. To be a refinement, the concrete design must resemble the abstract one, yet there must be significant differences between the two (otherwise, one of them is superfluous). Typically, the representation of data is different, the abstract design could for example be expressed using integers, whereas the concrete design uses bit-vectors. This relationship is captured by an abstraction function.

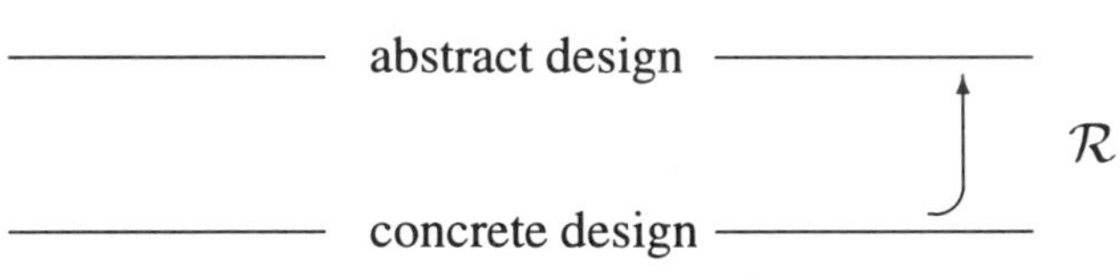

Formally, an abstraction function is a mapping from the concrete state space, S_c, to the abstract state space, S_a:

$$\mathcal{R} : S_c \mapsto S_a$$

Because the abstraction function maps concrete states to abstract states, it is possible that several different concrete states are mapped to a single abstract state. On the other hand, it is not required that the the concrete design can do everything that is allowed by the abstract. As a consequence of this, it is not required that the abstraction function is surjective, i.e., that there are concrete states mapping into every abstract state. Hence, there can be abstract states that have no counterpart in the concrete design (see also section 6.2.3).

If the abstract and concrete designs are complete cells, then the concrete must have exactly the same formal parameters as the abstract. Otherwise it is not possible to substitute the concrete design for the abstract. Furthermore, it is required that the abstraction function maps the formal parameters of the concrete design to the abstract one by one. This ensures that if the concrete cell does anything (visible to the environment), a similar change is observed in the abstract design.

Example: A Modulo-4 Counter

Consider a simple modulo-4 counter with an interface consisting of three state variables, $incr, mod4$, and ack. All three are of type boolean. When $incr$ becomes true, the modulo-4 counter must make either $mod4$ or ack true; $mod4$ becomes true once for every four times $incr$ is true, and ack becomes true the other three times. When $incr$ becomes false, both $mod4$ and ack must become false, before $incr$ is set to true again. This protocol can be described more precisely as follows:

PROTOCOL

$$mod4.pre \neq mod4.post \Rightarrow mod4.post = incr.pre \; AND$$
$$ack.pre \neq ack.post \Rightarrow ack.post = incr.pre \; AND$$
$$incr.pre \neq incr.post \Rightarrow incr.post = NOT(mod4.pre \; OR \; ack.pre)$$

An abstract design of the modulo-4 counter is shown in figure 6.2. The state of the counter is an integer c. In the concrete design, shown in figure 6.3, the integer is represented with two bits, (l, m); l is the least significant bit and m the most significant in the binary representation of c. The abstraction function relates the concrete state, s_c, consisting of l and m to the abstract state, s_a, consisting of c. Informally, the pair (l, m) is mapped to the integer $2 * m + l$.

```
CELL modulo-4(incr, mod4, ack: BOOLEAN)
  RESTRICTION mod4, ack: LOCAL
  STATE c: INTEGER
  INITIALLY c=0   mod4=FALSE   ack=FALSE
BEGIN
  ≪ incr AND NOT(mod4 OR ack) ≫ * (
    ≪ c=3 -> c, mod4:= 0, TRUE ≫ ||
    ≪ c≤2 -> c, ack:= c+1, TRUE ≫ ) ||
  ≪ NOT incr AND (mod4 OR ack) -> ack, mod4:= FALSE, FALSE ≫
END modulo-4
```

Figure 6.2 Abstract modulo-4 counter

Formally, this is specified as follows:

$$c.\mathcal{R}(s_c) = 2 * I(m.s_c) + I(l.s_c)$$

The indicator function $I\colon BOOLEAN \mapsto INTEGER$ converts a boolean to an integer ($I(TRUE) = 1$ and $I(FALSE) = 0$). If s_c is a concrete state then $m.s_c$ denotes the value of m in this state. Consider, for example, the concrete state, s_c, where $l.s_c = FALSE$ and $m.s_c = TRUE$.

$$\begin{aligned} c.\mathcal{R}(s_c) &= 2 * I(m.s_c) + I(l.s_c) \\ &= 2 * I(TRUE) + I(FALSE) \\ &= 2 \end{aligned}$$

So, the concrete state, $l.s_c = FALSE, m.s_c = TRUE$, is mapped to an abstract state where $c = 2$ as expected.

The formal parameters are also part of the state space, and hence the abstraction function also maps the concrete parameters to the abstract. This part of the mapping is rather trivial.

$$incr.\mathcal{R}(s_c) = incr.s_c, \quad mod4.\mathcal{R}(s_c) = mod4.s_c, \quad ack.\mathcal{R}(s_c) = ack.s_c,$$

the value of $incr$ in the concrete design is mapped to $incr$ in the abstract, $mod4$ in the concrete to $mod4$ in the abstract, and ack in the concrete to ack in the abstract. This part of the abstraction function is often not stated explicitly. However formally, the entire state space of the concrete design including parameters is mapped to the abstract state space. Section 6.4 contains a brief discussion of how to define refinement when the interfaces of the abstract and concrete designs are not identical.

```
CELL modulo-4(incr, mod4, ack: BOOLEAN)
  RESTRICTION mod4, ack: LOCAL
  STATE l, m: BOOLEAN
  INITIALLY l=FALSE m=FALSE mod4=FALSE ack=FALSE
BEGIN
  ≪ incr AND NOT(mod4 OR ack) ≫ * (
    ≪ NOT l -> l, ack:= TRUE, TRUE ≫ ||
    ≪ l AND NOT m -> l, m, ack:= FALSE, TRUE, TRUE ≫ ||
    ≪ l AND m -> l, m, mod4:= FALSE, FALSE, TRUE ≫ ) ||
  ≪ NOT incr AND (mod4 OR ack) -> ack, mod4:= FALSE, FALSE ≫
END modulo-4
```

Figure 6.3 Concrete modulo-4 counter

6.1.1 The Strong Refinement Condition

This section presents a sufficient condition for showing that a concrete design is a refinement. The condition is rather restrictive (strong), and it is based on the abstraction function. To show that a concrete design is a refinement, it must be shown that any behavior exhibited by the concrete design is also a possible behavior of the abstract design. This is ensured if the initial state of the concrete design is mapped to an initial state of the abstract, and if every execution of a concrete transition, t_c, has a corresponding abstract transition, t_a with a similar effect. This is formalized by the strong refinement condition.

Definition 6.2 *Two designs, D_a and D_c, meet the strong refinement condition, if and only if*

$$\forall t_c \in D_c : t_c(pre_c, post_c) \Rightarrow \exists t_a \in D_a : t_a(\mathcal{R}(pre_c), \mathcal{R}(post_c))$$

$$\mathcal{R}(init_c) \cong init_a$$

The first requirement is met if each transition in the concrete design, t_c, has a corresponding transition, t_a, in the abstract design, i.e., if t_c can change the state of the concrete design from pre_c to $post_c$, then t_a must change the abstract state from $\mathcal{R}(pre_c)$ to $\mathcal{R}(post_c)$. The second requirement forces the concrete variables to be initialized as prescribed by the abstract design. More precisely, the symbol $\cong$ means, all state variables, v, that are given an initial value in the abstract design must be given a corresponding initial value in the concrete design such that: $v.\mathcal{R}(init_c) = v.init_a$.

If the strong refinement condition is met, the concrete design is a refinement of the abstract. However, the condition is rather restrictive and relatively few designs meet it. Section 6.2 present a similar but weaker condition.

Example: The Modulo-4 Counter (continued)

In the modulo-4 counter, there are three transitions in the concrete design. Consider the first transition and assume that it changes the state, pre_c, to the state, $post_c$, then

$$incr.pre_c \ AND \ NOT(mod4.pre_c \ OR \ ack.pre_c) \ AND \ NOT l.pre_c$$

Furthermore, $l.post_c = TRUE$ and $m.pre_c = m.post_c$. Combining these, it can be concluded that:

$$\begin{aligned} c.\mathcal{R}(pre_c) &= 2 * I(m.pre_c) + I(l.pre_c) \\ &= 2 * I(m.pre_c) + I(FALSE) \\ &= 2 * I(m.pre_c) \\ &\leq 2 \end{aligned}$$

$$\begin{aligned} c.\mathcal{R}(post_c) &= 2 * I(m.post_c) + I(l.post_c) \\ &= 2 * I(m.post_c) + 1 \\ &= 2 * I(m.pre_c) + I(TRUE) \\ &= c.\mathcal{R}(pre_c) + 1 \end{aligned}$$

Hence, it follows that if t_c and t_a are the transitions:

$$t_c \ll incr \ AND \ NOT(mod4 \ OR \ ack) \ AND \ NOT l \rightarrow l, ack := TRUE, TRUE \gg$$

$$t_a \ll incr \ AND \ NOT(mod4 \ OR \ ack) \ AND \ c \leq 2 \rightarrow c, ack := c+1, TRUE \gg$$

then

$$t_c(pre_c, post_c) \Rightarrow t_a(\mathcal{R}(pre_c), \mathcal{R}(post_c))$$

which means that the strong refinement condition is met for the first transition. The reasoning for the other two transitions is similar. Hence, it can be shown that the abstraction function meets the strong refinement condition, and therefore, the concrete counter is a refinement of the abstract.

By using the strong refinement condition, it is possible to formally show that one design is a refinement of another. However, this refinement condition is very restrictive (strong), and hence only makes it feasible to show refinement for a limited class of realizations. Typically, a concrete design includes details that are not in the abstract design, for example, extra internal state variables such as the carry bits in an adder, or the internal state of a finite state machine. Changes in these are not reflected at the abstract level, which means that $pre_c \neq post_c$, but $\mathcal{R}(pre_c) = \mathcal{R}(post_c)$. When a concrete transition makes such an invisible state change the strong refinement condition is not met, because it requires that there is a corresponding abstract transition. This might not be the case when $\mathcal{R}(pre_c) = \mathcal{R}(post_c)$. In the next section, a weaker condition is presented.

6.2 THE WEAK REFINEMENT CONDITION

This section presents another sufficient condition for showing that a concrete design is a refinement. The condition is less restrictive (weaker) than the condition presented in section 6.1.1, which means that more designs will meet this condition. It is also based on the abstraction function as it was the case with the strong refinement condition. This condition is weakened in three respects: by allowing invisible state changes, by taking abstract invariants into account, and by taking concrete invariants into account.

6.2.1 Invisible State Changes

Usually, a concrete design contains more details than an abstract. Therefore, the abstraction function maps many concrete states to the same abstract state. Such redundancy can be necessary to achieve the desired efficiency or robustness in a realization. The strong refinement condition does not allow for this, because it requires that there is an abstract transition corresponding to *every* concrete one. Often in a redundant implementation, the concrete design makes state changes between two different concrete states representing the same abstract state, i.e., $\mathcal{R}(pre_c) = \mathcal{R}(post_c)$. Such changes are called invisible, because they are not reflected in the abstract design, and hence they do not prevent a concrete design from having the same behavior as an abstract design. To be a refinement, it is therefore sufficient to require that for each state change, $t_c(pre_c, post_c)$, there is *either* a corresponding abstract transition *or* that the state change is invisible.

$$\exists t_a \in D_a : t_a(\mathcal{R}(pre_c), \mathcal{R}(post_c)) \;\bigvee\; \mathcal{R}(pre_c) = \mathcal{R}(post_c)$$

As an example, consider an adder. At the abstract level, only the operands and the sum are relevant. The concrete design could be any of the adders considered in the previous chapters, e.g., the ripple carry adder or the Brent-Kung adder. In these adders, the carry is represented as an internal state variable, and hence, all transitions updating the carry are invisible, leaving the abstract state (where there is no carry) unchanged.

The requirement shown above is weaker than what is required in the strong refinement condition. It is, however, possible to weaken the condition even further by taking invariants into account as explained in the next section.

6.2.2 Utilizing Invariants

An invariant guarantees that a design only enters some subset of the state space. This can be utilized to weaken the refinement condition, because the concrete design needs only to behave like the abstract in the subset defined by the invariant, since these are the only states that will be entered during the computation. This makes it possible to show refinement in cases where it would otherwise not be possible (although the refinement condition will have more clauses, and therefore appear more complicated). Both invariants of the abstract and the concrete design can be utilized.

Invariants of the Abstract Design

An invariant of an abstract design also constrains all refinements; it defines a stable subset of the state space, i.e., these are the only states that are entered during the computation. Consider again two designs related by an abstraction function, $\mathcal{R}$.

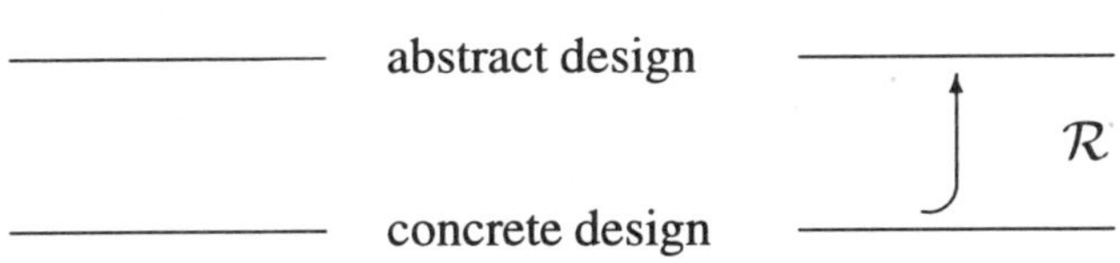

Assume that I_a is an invariant for the abstract design, and that $\mathcal{R}$ meets the strong refinement condition, then there is a corresponding invariant in the concrete design that is called an inherited invariant and is written as follows: $\mathcal{R}^{-1}(I_a)$. Note, that this is just a name of the invariant, there might not be a simple symbolic expression using the concrete state variables for $\mathcal{R}^{-1}(I_a)$. To check whether a concrete state, x_c, satisfies $\mathcal{R}^{-1}(I_a)$ one must check and see if $\mathcal{R}(x_c)$ satisfies I_a.

The notation, $I_a.s$, is used to denote the truth value of the invariant I_a in the state s, so, $\mathcal{R}^{-1}(I_a).x_c$ is the value of the invariant in state x_c; by definition $\mathcal{R}^{-1}(I_a).x_c \equiv I_a.\mathcal{R}(x_c)$.

If $\mathcal{R}$ *meets the strong refinement condition then* $\mathcal{R}^{-1}(I_a)$ *is an invariant for the concrete design.* To see this, assume that $\mathcal{R}^{-1}(I_a).pre_c$ holds, i.e., that $I_a.\mathcal{R}(pre_c)$ is satisfied, and that $t_c(pre_c, post_c)$. Then it follows (from the strong refinement condition) that:

$$\exists t_a : t_a(\mathcal{R}(pre_c), \mathcal{R}(post_c))$$

Since, I_a is an invariant for the abstract design, $I_a.\mathcal{R}(post_c)$ must hold, and hence, $\mathcal{R}^{-1}(I_a).post_c$. Therefore, $\mathcal{R}^{-1}(I_a)$ is an invariant for the concrete design.

The strong refinement condition states a requirement on all concrete states, $(pre_c, post_c)$, such that there is a transition, t_c, for which $t_c(pre_c, post_c)$. By utilizing an abstract invariant, I_a, this can be weakened by only considering concrete states that also satisfies $\mathcal{R}^{-1}(I_a)$. It is, therefore, only necessary to consider state pairs, $pre_c, post_c$, such that:

$$t_c(pre_c, post_c) \wedge \mathcal{R}^{-1}(I_a).pre_c$$

Invariants of the Concrete Design

In addition to the abstract invariants that are inherited, the concrete design may have additional invariants, for example, one constraining internal state variables.

Let I_c be a concrete invariant, then the refinement condition needs only to consider concrete states, $(pre_c, post_c)$ that satisfies I_c, i.e., $I_c.pre_c$.

$$t_c(pre_c, post_c) \wedge \mathcal{R}^{-1}(I_a).pre_c \wedge I_c.prec_c$$

By combining the weakenings discussed above, the weak refinement condition becomes:

Definition 6.3 *Two design, D_a and D_c, meet the strong refinement condition, if and only if*

$$\forall t_c \in D_c : t_c(pre_c, post_c) \wedge \mathcal{R}^{-1}(I_a).pre_c \wedge I_c.prec_c \Rightarrow$$
$$\exists t_a \in D_a : t_a(\mathcal{R}(pre_c), \mathcal{R}(post_c)) \ \vee \ \mathcal{R}(pre_c) = \mathcal{R}(post_c)$$

$$\mathcal{R}(init_c) \cong init_a$$

It must be stressed that the weak refinement condition is only a sufficient condition. However, it is formulated to allow for a verification by many (small) cases; one for each transition in the concrete design. This is similar to the verification technique for invariants and protocols, which is also done by considering each transition separately. In section 6.3 it is shown, how the refinement verification can be mechanized in the same way as design verification. It is possible to formulate weaker conditions, and a lot of theoretical work is concerned with finding sufficient and necessary (i.e., the weakest possible) conditions, however, such a condition may not be practically useful. For practical verification, the strength of the condition is only one aspect; the ease of use and possibilities for mechanical support are equally important aspects. This has clearly influenced the formulation of the weak refinement condition shown above.

Example: A Modulo-5 Counter

Figure 6.4 shows a simple modulo-5 counter similar to the modulo-4 counter described in section 6.1. The only differences to the modulo-4 counter are the substitutions of 4 by 5, etc., and the addition of an invariant $c < 5$. It is quite simple to demonstrate that this is an invariant and the details are omitted here.

To realize this counter three bits are needed in the concrete design, (l, o, m). l is the least significant bit and m the most significant in the binary representation of c. The abstraction function is:

$$c.\mathcal{R}(s_c) = 4 * I(m.s_c) + 2 * I(o.s_c) + I(l.s_c)$$

It is not possible to show that the abstraction function (and that the two designs) meet the strong refinement condition. Consider, for example, the concrete state: $m.pre_c = TRUE$ and $o.pre_c = TRUE$. From this state, the first transition leads to a post state that does not have an abstract counterpart. However, by taking the abstract invariant, $c < 5$ into account, it becomes clear that the concrete design never enters the state where $m.pre_c = TRUE$ and $o.pre_c = TRUE$. By utilizing the invariant, it is possible to show that the weak abstraction condition

```
CELL modulo-5(incr, mod5, ack: BOOLEAN)
  RESTRICTION mod5, ack: LOCAL
  STATE c: INTEGER
  INVARIANT c<5
  INITIALLY c=0  mod5=FALSE  ack=FALSE
BEGIN
   ≪ incr  AND  NOT(mod5  OR  ack) ≫ ∗ (
      ≪ c=4  –>  c, mod5:= 0, TRUE ≫ ||
      ≪ c≤3  –>  c, ack:= c+1, TRUE ≫  ) ||
   ≪ NOT incr  AND  (mod5  OR  ack)  –>  ack, mod5:= FALSE, FALSE ≫
END modulo-5
```

Figure 6.4 Abstract modulo-5 counter

```
CELL modulo-5(incr, mod5, ack: BOOLEAN)
RESTRICTION mod5, ack: LOCAL
STATE l,o,m: BOOLEAN
INITIALLY l=FALSE o= FALSE m=FALSE mod5=FALSE ack=FALSE
BEGIN
   ≪ incr  AND  NOT(mod5  OR  ack) ≫ ∗ (
   ≪ NOT l  AND  NOT m   –>  l, ack:= TRUE, TRUE ≫ ||
   ≪ l  AND  NOT o  –>  l, o, ack:= FALSE, TRUE, TRUE ≫ ||
   ≪ l  AND  o   –>  l, o, m, ack:= FALSE, FALSE, TRUE, TRUE ≫||
   ≪ m  –>  m, mod5:= FALSE, TRUE ≫) ||
   ≪ NOT incr  AND  (mod5  OR  ack)  –>  ack, mod5:= FALSE, FALSE ≫
END modulo-5
```

Figure 6.5 Concrete modulo-5 counter

```
CELL fifo(in, out: element)
  STATE q: SEQUENCE OF element
BEGIN
  ≪ in ≠ E –> q, in:= q⊙in, E ≫ ||
  ≪ out=E AND q=e⊙r –> q, out:= r, e ≫
END fifo
```

Figure 6.6 Abstract FIFO design

is met. The arguments are very similar to those given for the modulo-4 counter, and are therefore not repeated. It can be concluded that the concrete design is a refinement of the abstract.

Example: A FIFO Queue

Consider again the FIFO queue described in section 2.7. It is a data structure that holds a sequence of elements. New elements are added to one end of the sequence and removed from the other end. The interface to the FIFO consists of two registers, an input register, in and an output register out. The static E (empty) denotes a value that is different from all values contained in the queue. An element can be inserted in the queue when the input register is empty. Insertion is done by assigning an element to in. An element can be removed when the output register is full ($out \neq E$). The removal is done by changing the value of out to E. Figure 6.6 shows a design description of an abstract FIFO queue. This abstract design uses the data type $SEQUENCE$ that has a concatenation operator $\odot$. If q is of type $SEQUENCE$, then $q \odot in$ denotes the sequence q concatenated with the value of in. In the concrete design, shown in figure 6.7, the sequence is realized with an array. The abstraction function, $\mathcal{R}$, maps the array used in the concrete design into a sequence. An auxiliary function F is used to express the abstraction function.

$$q.\mathcal{R}(s_c) = F(q.s_c, n-1)$$

where

$$F(q,i) = \begin{cases} \lambda & \text{if} \quad i < 0 \\ F(q, i-1) & \text{if} \quad q[i] = E \\ q[i] \odot F(q, i-1) & \text{if} \quad q[i].s \neq E \end{cases}$$

λ is the empty sequence. As it has been the case with the other examples considered in this chapter, it is implicitly assumed that the formal parameters

```
CELL fifo(in, out: element)
   STATE q: ARRAY[0..n-1] OF element
   TRANSITION transfer(from, to: element)
      ≪ from ≠ E  AND  to=E  ->  to, from:= from, E ≫
BEGIN
   transfer(in, q[0])  ||  transfer(q[n-1], out) ||
   {|| i: [0..n-2]  |  transfer(q[i], q[i+1]) }
END fifo
```

Figure 6.7 Concrete FIFO design

of the concrete design are mapped to the formal parameters of the abstract design.

It is now shown that $\mathcal{R}$ meets the weak refinement condition. There are three cases (three kinds of concrete transitions) to consider.

1. $t_{transfer(in,q[0])}(pre_c, post_c)$

   From the transition description it follows that:

   $in.pre_c \neq E, q[0].pre_c = E, in.post_c = E, q[0].post_c = in.pre_c.$

$$\begin{aligned} q.\mathcal{R}(post_c) &= F(q.post_c, n-1) \\ &= F(q.pre_c, n-1) \odot q[0].post_c \\ &= q.\mathcal{R}(pre_c) \odot q[0].post_c \\ &= q.\mathcal{R}(pre_c) \odot in.pre_c \end{aligned}$$

   Hence, it follows that from the pre-state $\mathcal{R}(pre_c)$ the abstract transition:

   $\ll in \neq E \; \text{->} \; q, in := q \odot in, E \gg$

   leads to the post-state $\mathcal{R}(post_c)$.

2. $t_{transfer(q[i],q[i+1])}(pre_c, post_c)$

   From the transition description it follows that:

   $q[i].pre_c \neq E, q[i+1].pre_c = E$. Therefore,

$$\begin{aligned} q.\mathcal{R}(post_c) &= F(q.post_c, n-1) \\ &= F(q.pre_c, n-1) \\ &= q.\mathcal{R}(pre_c) \end{aligned}$$

Furthermore, none of the formal parameters in, out are changed. Hence, this concrete transition makes an invisible state change.

3. The third transition (output) is similar to the first.

Note that step 2 (the transfer transitions) is an example of a transition making an invisible state change. From these three cases, it follows that $\mathcal{R}$ meets the refinement condition.

Example: Brent-Kung Adder

The derivation of the Brent-Kung adder from the Chandy-Misra adder (see section 4.3.3) is an example of a refinement. The abstract design uses (the Chandy-Misra adder) a three valued type, $(U, 1, 0)$, and in the concrete design this is represented by a pair of booleans (p, g), defined by the following table:

$p[i]$	$g[i]$	$d[i]$
0	0	0
0	1	1
1	-	U

The table defines an abstraction function, mapping the concrete state space, represented by (p, g) to the abstract state space, represented by d_a.

$$d_a[i].\mathcal{R}(s_c) = \begin{cases} U & \text{if} \quad s_c.p[i] = 1 \\ s_c.g[i] & \text{if} \quad s_c.p[i] = 0 \end{cases}$$

The abstract design of the carry computation is described as follows (the update of t is ignored here):

$$\{+i : [1..n] \mid \ll d_a[i] = U \rightarrow d_a[i] := d_a[i-t] \gg \}$$

In the concrete design, the carry computation is done as follows:

$$\begin{aligned} &\{+i : [1..n] \mid \\ &\quad \ll d_c[i].p := d_c[i].p \ AND \ d_c[i-t].p \gg + \\ &\quad \ll d_c[i].g := d_c[i].g \ OR \ (d_c[i].p \ AND \ d_c[i-t].g) \gg \} \end{aligned}$$

It can be verified that this is a refinement, by showing that the abstraction function meets the (weak) refinement condition. This is not possible unless the following invariant of the concrete design is taken into account:

$$\mathit{INVARIANT} \quad \mathit{FORALL}\ i : [1..n] \ . \ \ d_c[i].p \Rightarrow NOT\ d_c[i].g$$

In problem 4.10 it was shown that the Brent-Kung adder does indeed maintain this invariant.

To show that the weak refinement condition is met, assume that pre_c is a concrete state where the invariant holds and that $post_c$ is the corresponding post-state (note that this is a synchronous design, therefore, all transitions are executed simultaneously). Consider a single element of d_c, $d_c[i]$. The following cases cover all possible state changes.

$(d_c[i].p).pre_c = TRUE \wedge (d_c[i-t].p).pre_c = TRUE$

In the concrete design the assignment $d_c[i].p := d_c[i].p\ AND\ d_c[i-t].p$ results in a post-state where $(d_c[i].p).post_c = TRUE$. In the abstract design, $d_a[i].\mathcal{R}(pre_c) = U$, $d_a[i-t].\mathcal{R}(pre_c) = U$, and $d_a[i].\mathcal{R}(post_c) = U$. Therefore, the abstract transition $d_a[i] := d_a[i{-}t]$, satisfies $t_a(\mathcal{R}(pre_c), \mathcal{R}(post_c))$, so the refinement condition is met in this case.

$(d_c[i].p).pre_c = TRUE \wedge (d_c[i-t].p).pre_c = FALSE$

In the concrete design the assignment $d_c[i].p := d_c[i].p\ AND\ d_c[i-t].p$ results in a post-state where $(d_c[i].p).post_c = FALSE$. From the invariant it follows that $(d_c[i].g).pre_c = FALSE$ (because $(d_c[i].p).pre_c = TRUE$), hence, the assignment $d_c[i].g := d_c[i].g\ OR\ (d_c[i].p\ AND\ d_c[i-t].g)$ results in a results in a post-state where $d_c[i].post_c = d_c[i-t].pre_c$. When $(d_c[i].p).pre_c = TRUE$, then $d_a[i].\mathcal{R}(pre_c) = U$. Therefore, doing the assignment $d_a[i] := d_a[i-t]$ in the abstract design corresponds to what happens in the concrete design, where it was concluded that $d_c[i].post_c = d_c[i-t].pre_c$. It follows, that the refinement condition is met in this case.

$(d_c[i].p).pre_c = FALSE \wedge (d_c[i-t].p).pre_c = TRUE$

In this case neither $d_c[i]$ nor $d_a[i]$ is changed, so the refinement condition is met.

$(d_c[i].p).pre_c = FALSE \wedge (d_c[i-t].p).pre_c = false$

In this case neither $d_c[i]$ nor $d_a[i]$ is changed, so the refinement condition is met.

6.2.3 Further Restrictions on the Abstraction Function

The refinement condition allows for very significant differences between the abstract and the concrete designs. Consider, for example, the FIFO where the abstract design allows for an unbounded number of elements in the FIFO. In contrast, the concrete design has a fixed capacity of n elements. In some cases this latitude is desirable, but there are cases where further restrictions on the abstraction function are needed. One could, for example, require that the abstraction function is surjective (every abstract state is represented by at least one concrete state). Such a requirement would not allow a concrete design with fixed capacity to represent an abstract one with unbounded capacity. This is a research topic which currently receives significant attention, but it is not pursued in further depth here.

6.3 MECHANIZATION

As is the case with verifying invariants, most refinement proofs are quite tedious. Therefore, it is important to find a way of mechanizing the laborious parts of the verification. This section describes how a translator and the mechanical theorem prover LP can be used to check the refinement condition in a way very similar to the one used for verifying invariants and protocols.

States and transitions are formalized as described in section 3.3.3. The pre- and post states are called *pre* and *post* respectively and a transition, t, such as

$$\textit{TRANSITION} \quad t \quad \ll c -> v := e \gg$$

is formalized as the equations:

$$t(pre, post) == c.pre \wedge v.post = e.pre \wedge U(\ldots)$$

where ... are all other variables than v; these are unchanged.

As an example consider the first transition of the modulo-5 counter:

$$\begin{aligned} t_{c1}(pre, post) == {} & ((incr.pre \wedge NOT(mod5.pre \vee ack.pre) \ AND \ c.pre = 4) \ AND \\ & c.post = 0 \ AND \ mod5.post = TRUE \ AND \\ & incr.post = incr.pre \ AND \ ack.post = ack.pre \ ) \end{aligned}$$

A similar formalization is needed for all the other transitions of both the abstract and concrete design. This is a simple syntactical translation that can be automated.

To check the refinement condition, the abstraction function must also be provided explicitly and expressed in a way that can be handled by the theorem prover. For the modulo-5 counter this is quite simple using the following functions.

$$I(true) == 1 \quad I(false) == 0$$
$$R(s_c) == 4 * I(m.s_c) + 2 * I(o.s_c) + I(l.s_c)$$

Using these definitions, the refinement condition is verified by showing an implication of the following form, for each of the the concrete transitions, for example t_{c1}:

$$\begin{aligned} &\textit{prove } t_{c1}(pre, post) \textit{ AND } Inv(pre) \Rightarrow \\ &\qquad t_{a1}(R(pre), R(post)) \lor t_{a2}(R(pre), R(post)) \lor t_{a3}(R(pre), R(post) \lor \\ &\qquad R(pre) = R(post) \end{aligned}$$

In simple cases like the counter, LP can prove such implications automatically, while in other cases some user assistance is needed.

There is a prototype translator that generates the necessary verification conditions for showing the weak refinement condition, given two designs and an abstraction function. This translator is based on ST2LP, and it has been used to verify examples such as the ones shown in this chapter.

6.4 INTERFACE REFINEMENT

By defining refinement as "subset of behavior" one gets the nice property, that in any environment it is possible to replace an abstract design by one of its refinements. If it is a refinement, then the concrete design only exhibits behaviors that the abstract design also have, hence, the environment cannot tell any difference between the two. This is a nice property, however, in many cases it seems too restrictive to capture the notion of a correct implementation. Consider, for example, the abstract and concrete designs outlined below (not all internal details are shown).

CELL $abstract\_adder(a, b, s$:*INTEGER*)
BEGIN $\ll s := a + b \gg$ *END* $abstract\_adder$
TYPE $word$ = *ARRAY [0..15] OF BOOLEAN*
CELL $concrete\_adder(a, b, s : word)$
BEGIN ... *END* $concrete\_adder$

In these designs, the type of the operands in the abstract and concrete designs are not the same, hence, it is not possible to directly replace the abstract with the concrete design. The problem is that the interfaces of the two designs are different, first of all, they have different types, but the behavior at the interface is also different. In the concrete design, the individual bits (array elements) of the sum s may change one at a time. Hence, the environment might observe a number of intermediate results before the result, s, stabilizes at the correct sum. This will not happen in the abstract design where there is only one transition giving the correct sum in s directly.

To be able to verify refinements, where there is some change of interface, it is necessary to modify the definition. It can still be based on the idea of subsets, but computations (sequences of states) should be replaced by another concept capturing behavior, where differences in the interface is allowed. This is currently an active research topic and beyond the scope of this book.

6.5 BACKGROUND

To verify refinement requires that both an abstract and a concrete design have been completed. In chapter 3 this type of verification is called implementation verification to distinguish it from design verification where only one design, the abstract, is needed. Theoretically, it is not necessary to distinguish the two types of verification. The invariants and protocols can be interpreted as an abstract design by themselves, and verifying that they are consistent with a design description is done as a refinement proof. From a theoretical point of view, this has the advantage of unifying the two types of verification. However, in this book they are treated separately to simplify the presentation and the verification techniques.

Refinement has been discussed extensively in the research literature. The notion of abstraction functions was proposed in 1972 by Hoare [34]. At that time the emphasis was on refining the data representation of a computation. The definition of refinement used here (subset of behaviors) stems from [43]. However, this work and most of the recent research address the more complicated issue of showing refinements that preserve both safety and liveness properties whereas this chapter (like the rest of the book) is concerned with safety only (see section 3.5). The argument for concentrating on safety properties is mainly pragmatic, namely that this is already a difficult verification task both in terms of the effort required by the designer and by the mechanical tools. Maybe it is only a first step, but a necessary first step which must be taken before even more involved and demanding verification tasks are attempted. This is discussed further in the epilog of the book.

The use of abstraction functions (or something similar) is a very common approach to verifying refinement. In 1988 it was shown that under suitable (and reasonable) assumptions, then it is always possible to find an abstraction function [44] (if the concrete design is a refinement the abstract). In other words, it is possible to weaken the refinement conditions even further than what is shown here, to obtain a necessary and sufficient condition.

There are many mechanical tools available for implementation verification. As is the case for verifying invariants and protocols, both model checkers and theorem provers are used. The survey paper [32] contains a nice overview of the various approaches that has been used for circuit verification.

6.6 EXERCISES

Exercise 6.1 Consider the following abstract and concrete adder designs.

$CELL\ a(l, f : BOOLEAN;\ x, y, s : INTEGER)$
$BEGIN$
$\quad \ll l\ AND\ NOT f \rightarrow l := FALSE \gg ||$
$\quad \ll NOT f \rightarrow s := 0 \gg ||$
$\quad \ll NOT l\ AND\ s = x + y \rightarrow f := TRUE \gg$
$END\ a$

$CELL\ c(l, f : BOOLEAN;\ x, y, s : INTEGER)$
$\quad STATE\ xl, yl : INTEGER$
$\quad INVARIANT\ NOT l \Rightarrow (s + xl + yl = x + y)$
$BEGIN$
$\quad \ll s = 0\ AND\ l\ AND\ NOT f \rightarrow xl, yl, l := x, y, FALSE \gg ||$
$\quad \ll NOT f\ AND\ l \rightarrow s := 0 \gg ||$
$\quad \ll xl = 0\ AND\ yl = 0\ AND\ NOT l \rightarrow f := TRUE \gg ||$
$\quad \ll NOT f\ AND\ xl > 0\ AND\ NOT l \rightarrow s, xl := s + 1, xl - 1 \gg ||$
$\quad \ll NOT f\ AND\ yl > 0\ AND\ NOT l \rightarrow s, yl := s + 1, yl - 1 \gg$
$END\ c$

Find an abstraction function and use it to show that the concrete design, c is a refinement of the abstract design, a.

Exercise 6.2 Verify that $c < 5$ is an invariant for the design shown in figure 6.4.

Exercise 6.3 Consider these transitions taken from the Brent-Kung adder

$$\{+i : [1..n] \mid \ll d_c[i].p := d_c[i].p\ AND\ d_c[i-t].p \gg + \\ \ll d_c[i].g := d_c[i].g\ OR\ (d_c[i].p\ AND\ d_c[i-t].g) \gg\}$$

Can the last transition be changed to:

$$\ll d_c[i].g := d_c[i].g\ OR\ d_c[i-t].g \gg$$

Try to answer the question by showing that the changed design is a refinement of the abstract design given in section 6.2.2.

Exercise 6.4 Consider the following abstract and concrete designs.

```
TYPE element = RECORD t, f : BOOLEAN END
CELL a(in, out, succ : element)
  FUNCTION E(e : element) : BOOLEAN
    RETURN NOTe.f AND NOTe.t
BEGIN
  ≪ (in = E)≠(succ = E) -> out := in ≫
END a

CELL c(in, out, succ : element)
  STATE temp : BOOLEAN
  FUNCTION dual(x : element) : BOOLEAN
    RETURN NOT(x.t AND x.f)
 INVARIANT
   dual(in) AND dual(out) AND dual(succ) AND
   (((out.f OR out.t) AND (in.f OR in.t)) ⇒
        ((out.f = in.f) AND (out.t = in.t)))
BEGIN
  ≪ in.f <=> NOT(succ.t OR succ.f) -> out.f := in.f ≫ ||
  ≪ in.t <=> NOT(succ.t OR succ.f) -> out.t := in.t ≫
END c
```

Find an abstraction function, and use it to show that the concrete design, c, is a refinement of the abstract design a.

Exercise 6.5 Consider an abstract modulo-n counter:

```
CELL modulo_n(incr, modn, ack : BOOLEAN; STATIC n : INTEGER)
  RESTRICTION modn, ack : LOCAL
  STATE c : INTEGER
  INITIALLY c = 0  modn = FALSE  ack = FALSE
BEGIN
  ≪ incr AND NOT(modn OR ack) ≫ * (
      ≪ c = n−1 -> c, modn := 0, TRUE ≫ ||
      ≪ c ≤ n−2 -> c, ack := c + 1, TRUE ≫ ) ||
  ≪ NOTincr AND (modn OR ack) -> ack, modn := FALSE, FALSE ≫
END modulo_n
```

Describe a concrete design of the modulo-n counter using only state variables of type boolean. Show that the concrete design is a refinement of the abstract.

Exercise 6.6 Consider the following abstract and concrete oscillator designs.

```
CELL a
  STATE x, y, z : BOOLEAN
  INITIALLY x = TRUE  y = FALSE  z = TRUE
BEGIN
   << x≠y -> z := y >> ||
   << x := z >> ||
   << y := NOT z >>
END a
CELL c
  STATE x, y, z, zz : BOOLEAN
  INITIALLY x = TRUE  y = FALSE  z = TRUE
BEGIN
   << x≠y -> z := y >> || << zz := z >> ||
   << x := zz >> ||
   << y := NOT z >>
END c
```

Use the weak refinement condition to show that the concrete design, c, is a refinement of the abstract design, a.

7

SELF-TIMED CIRCUITS

Circuits that operate correctly without relying on global synchronization are called self-timed. The absence of global synchronization prevents the use of signals from a global clock which are assumed to arrive (almost) simultaneously at all subcircuits. Instead, the progress of computation is controlled by the values of communication signals. Consider the design shown in figure 7.1. Whenever x and y have different values – for example, in the initial state – z gets the value of y that enables the transitions changing x and y. After both have changed, their values are again different, leading to a new change of z, etc. No matter what the delays are in the computation of the gates and in the propagation of values along wires carrying x, y, and z, the values of the three state variables oscillate. It is not possible to find a combination of delays that makes the oscillation stop, so it is an example of a self-timed design.

```
CELL oscillator
  STATE x, y, z: BOOLEAN
  INITIALLY x=TRUE  y=FALSE  z=TRUE
BEGIN
  ≪ x≠y -> z:= y ≫ ||
  ≪ x:= z ≫ ||
  ≪ y:= NOT z ≫
END oscillator
```

Figure 7.1 Oscillator

Figure 7.2 Human pipeline

This simple example is merely shown to illustrate how a self-timed circuit operates. Later in this chapter, more interesting designs are discussed, for example, a FIFO queue and a modulo-n counter.

Self-timed circuits are data driven, and communication between components is based on handshake protocols. Therefore, it is possible to replace part of a design, e.g., change performance characteristics, without any concern for other parts of the design. This composition ability is an important advantage of self-timed circuits. In contrast, large synchronous designs tend to resist incremental improvements, because the entire design must be adjusted when changes in one part require a change of the clock frequency. Similarly, it is likely that redesign of one part influences a carefully designed clock buffering and distribution scheme, thus affecting many other parts of the design – a task that requires detailed knowledge of the clock load and clock delay in all other components.

The following example illustrates some of the dangers one may encounter in a synchronous system.[1] A truckload of watermelons has to be moved from a truck to a warehouse. This is done by arranging a number of workers into a pipeline going from the truck to the warehouse (see figure 7.2). The watermelons are removed from the truck by the first person in the pipeline. He throws them to the second person, one by one; the second person passes on the watermelons to the third, etc. The foreman of the workers has decided to

[1] The example was suggested by Dr. Michael Kishinevsky, who called it a Russian pipeline

make the pipeline synchronous, so at regular intervals he shouts "throw" and everybody throws the watermelon he is currently holding to the next person in the pipeline.

This example illustrates how vulnerable a synchronous system can be. It is, for example, easy to imagine how disastrous it would be if the "throw" signal from the foreman does not arrive exactly at the same time to all workers, or if their reaction times are very different. The example also illustrates the difficulty of constructing synchronous systems that are spread over large geographical areas. This is in contrast to self-timed circuits that are synchronized using local handshakes, for example, by having one worker shout "ready" to his neighbor when he is prepared to receive the next watermelon. The handshaking makes self-timed computations very adaptable, and insensitive to fluctuations caused by operating conditions, data, or fabrication variations.

Many claims have been made about possible advantages and disadvantages of self-timed circuits, for example, that they consume less power and run faster than corresponding synchronous circuits. This discussion is not pursued here; the main emphasis is on showing how Synchronized Transitions is used to design self-timed circuits, and in particular, how to use a design description to verify that a self-timed realization is possible. This enables the designer to use the high-level description of a design to detect if there are parts that cannot be realized with self-timed circuitry. Before presenting the verification technique, some important subclasses of self-timed circuits are defined.

7.1 CLASSIFICATION

Self-timed circuits cover a range of different circuit families, for example:

- Circuits controlled by distinct local clocks, but interconnected asynchronously.
- Circuits operating without any clocks, but based on delay assumptions about gates and wires.
- Circuits operating correctly regardless of the magnitude of delays caused by gates, but assuming no delays in wires.
- Circuits operating correctly regardless of the magnitude of delays caused by gates and wires.

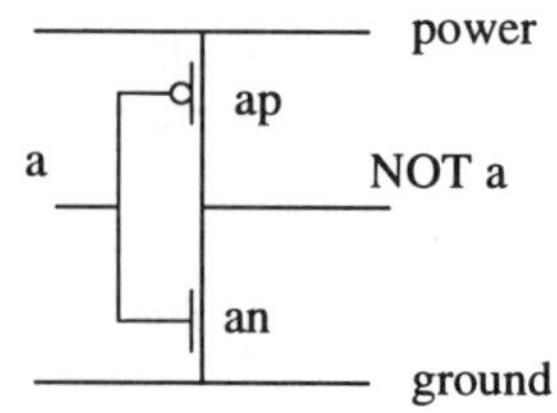

Figure 7.3 CMOS inverter

Other families are studied and used in various applications. The main differences between the families are the assumptions made about delays to guarantee correct operation. The ultimate would be not to make any assumptions at all. Circuits in this ultimate family must operate correctly no matter what the magnitude is of any delay anywhere in the circuit. From a practical point of view, this ultimate class is not very interesting. Consider, for example, a simple CMOS inverter as shown in figure 7.3. This inverter only works correctly if the forking wire, distributing the input, a, to the gates of the n- and p-transistors, ensures that the same signal is present at both gates. If the wire leading to the p-transistor causes a big delay of the input, then it is possible to have a computation where both transistors are conducting, which effectively is a short-circuit. When this happens, it is unlikely that the circuit continues to operate correctly. Hence, it is necessary to assume that the signal present at the two gates, ap and an, is the same. This simple example shows that any interesting family of self-timed circuits must be based on some delay assumptions, otherwise it will only contain trivial and uninteresting circuits, it would for example not even contain an inverter.

The inverter example illustrates why most practical interest is aimed at families of self-timed circuits based on some delay assumptions, for example, that there are no delays in wires, or that the delays in certain explicitly mentioned wires are bounded. The exact choice of delay assumptions made in a particular design depends on many factors like technology, operating conditions, and environment. It is a typical engineering compromise between many opposing factors. In this chapter, the focus is on the following two families:

- **Speed-independent:** Circuit elements are allowed to have arbitrary, positive, finite delays, but wires are assumed to have no delays.
- **Delay-insensitive:** Both circuit elements and wires may have arbitrary, positive, finite delays, but assumptions about wire delays *within* circuit elements are allowed.

```
CELL oscillator
  STATE x, y, z, stop: BOOLEAN
  INITIALLY x=FALSE y=TRUE z=FALSE stop=FALSE
BEGIN
  ≪ x ≠ y –> z:= y ≫ ||
  ≪ y ≠ z –> x:= z ≫ ||
  ≪ z ≠ x –> y:= x ≫ ||
  ≪ (x=y) AND (y=z) –> stop:= TRUE ≫
END oscillator
```

Figure 7.4 A speed-independent oscillator

A circuit element is a building block, for example, fundamental gates such as inverters and NAND-gates. It could also be a more complex gate such as an adder. The exact functionality of the circuit elements is not so important for the techniques introduced in this chapter. It is, however, important to identify a level of primitives whose internal properties may depend on local delay assumptions.

Both families listed above are defined as properties of the circuit realization. It would be impractical if one had to construct a circuit realization before determining whether a design is speed-independent or delay-insensitive. Section 7.3.1 shows an implementation condition for a design in Synchronized Transitions that ensures that a speed-independent realization is possible, and section 7.5 discusses how to ensure delay-insensitivity.

Example: An Oscillator

Figure 7.4 shows an oscillator design. The three state variables x, y, and z alternate between true and false. It is not difficult to verify that the assertion $NOT((x = y)\ AND\ (y = z))\ AND\ NOT stop$ is an invariant for this design. Hence, the last transition is never activated in a correct realization of the design.

This oscillator is speed-independent, because no delay of any of the transitions can change its behavior, i.e., stop the oscillation or enter a state violating the invariant. The first three transitions are very similar, and when one of them has executed, it cannot execute again until both of the other transitions have finished their next execution. In section 7.3.1 it is shown how this claim can be verified formally.

The oscillator is speed-independent, but it is not delay-insensitive. It can malfunction if there are different delays in wires carrying the value of the state variables. Consider a state where

$$(x, y, z) = (\mathit{FALSE}, \mathit{TRUE}, \mathit{FALSE})$$

Assume that the wire leading y to the last transition is very slow, so the true value of y remains at the input of the last transition even after the value of y has changed on the other wires connecting the first three transitions. After a while, it is possible that the last transition observes a state where

$$(x, y, z) = (\mathit{TRUE}, \mathit{TRUE}, \mathit{TRUE})$$

In this state, the last transition is enabled, and performing it makes the state variable stop true. Hence, the oscillator exhibits a different behavior because of the wire delays, and therefore it is not delay-insensitive although it is speed-independent.

In section 7.5 delay-insensitivity is defined formally, and it is discussed how it can be verified that a design is delay-insensitive.

7.2 MODELS OF SELF-TIMED CIRCUITS

In the computational model of Synchronized Transitions transitions are assumed to be atomic. This allows the designer to construct, analyze, and verify designs by considering each transition separately. Unfortunately, this model does not reflect how circuits operate. Many subcircuits may change their outputs simultaneously, and a change made by one subcircuit can interfere with the computations of another. Furthermore, in a physical circuit, the change from one signal value to another is not instantaneous. During a change, signals may assume values that are not well-defined discrete values. For example, a signal corresponding to a boolean state variable could have a value that some subcircuits interpret as true and others as false. In synchronous circuits, such problems are avoided by using the clock signal to indicate when the physical signals have well-defined values corresponding to the discrete values used in the abstract models. A similar correspondence must be established in a self-timed circuit.

It is beyond the scope of this book to provide a complete justification of the consistency between an abstract (discrete) model and a physical realization. In general, the correspondence between the physical reality and a model cannot be established by formal, rigorous means, since such methods of reasoning do not apply to physical objects and quantities. Instead, physical experiments and observations must be used. Furthermore, there is quite a big gap between physical circuits and the computational model of designs in Synchronized Transitions; to relate the two, several intermediate models are needed. In section 7.3, such an intermediate model is presented and related to the computational model. The intermediate model is discrete, but reflects delays in gates and wires.

The computational model of Synchronized Transitions allows active transitions to be executed in any order. This nondeterminism appears as an arbitrary delay between satisfying the precondition of a transition and the subsequent execution. The nondeterminism corresponds to the gate delays in a circuit that realize the transition. Hence, the computational model of Synchronized Transitions captures gate delays directly. If, for example, it is verified that a design maintains a certain invariant, I, it means that I holds regardless of the order in which transitions are executed, which in turns ensures that I holds regardless of gate delays. However, this assumes that the atomicity is preserved by the realization. A similar assumption is needed to ensure correct operation of the synchronous two-phase realization described in section 5.1. In the synchronous case, an implementation condition CREW is defined; this condition is a sufficient condition to ensure that the two-phase realization corresponds to an atomic execution. A similar approach is used here for the self-timed realizations, where an implementation condition, called stability, is defined. It is shown that this condition is sufficient to ensure that a speed-independent realization only enters states that can also be entered by an atomic execution. Low-level circuit models are beyond the scope of this book and, therefore, concrete circuits are not shown.

Section 7.5 discusses how wire delays can be modeled and how it can be verified that a design is delay-insensitive.

7.3 SPEED-INDEPENDENCE

This section contains a formal characterization of speed-independence. It is very similar to the classical definition given by Muller and Bartkey [54] in their pioneering work on self-timing. The formal characterization is used to formulate an implementation condition, called stability, which allows the designer to check that a design allows for a speed-independent realization. The stability condition is a sufficient condition formulated in such a way that it can be checked mechanically in a manner similar to the way invariants, protocols, and the implementation condition CREW are checked.

As the name indicates, speed-independence implies that the overall behavior of a design does not depend on the computation times of its individual parts. The model introduced below is called the Muller model, after David Muller [54], who used the model to formulate a theory of speed-independent circuits.

A circuit consists of a set of boolean variables, $(z_1, z_2, \ldots, z_n)$ $(n \geq 1)$, each of which has an associated boolean function:

$$f_i(z_1, z_2, \ldots, z_n) \mapsto boolean \quad 1 \leq i \leq n$$

Intuitively, the z variables correspond to the outputs of circuit elements, and f_i, specifies how to compute a new value of z_i, called z_i', based on the current values of all outputs, $(z_1, z_2, \ldots, z_n)$:

$$z_i' = f_i(z_1, z_2, \ldots, z_n)$$

The state of a circuit is an n-tuple of boolean values, one for each output of a circuit element. When $z_i' \neq z_i$, the circuit is in a state where the value of z_i can change.

Definition 7.1 *The variable z_i is excited if and only if $z_i' \neq z_i$.*

Definition 7.2 *A state $(z_1, z_2, \ldots, z_n)$ is a conflict state if and only if there exists two different excited variables z_i, z_j such that*

$$z_i = f_i(z_1, z_2, \ldots, z_j', \ldots, z_n) \ \ or \ \ z_j = f_j(z_1, z_2, \ldots, z_i', \ldots, z_n)$$

Intuitively, a conflict state is a state where changing an excited variables will make another variable not excited. Conflict states indicate that the circuit is not speed-independent, because the timing of the changes of the excited variables can affect the behavior.

Definition 7.3 *A circuit with no conflict states is speed-independent.*

In Muller's original work, it was shown how this property (called semimodularity) was sufficient to ensure that the behavior of a circuit is independent of computation speeds. The reader is referred to the original writings of Muller [54] for a more thorough discussion of the definition.

Often, it is not possible to show that a design is speed-independent without taking some invariant into account. Such designs only exhibit speed-independent behavior if they are initialized in a state satisfying the invariant.

7.3.1 The Implementation Condition Stability

The definition of speed-independence given above does not directly suggest a practical procedure for checking that a design is speed-independent. Below, an implementation condition, called stability, is formulated. This condition is (relatively) simple to check, but it is only a sufficient and not a necessary condition. Hence, there are speed-independent designs that do not meet the condition. This might seem unsatisfactory from a theoretical point of view, but hopefully, the examples given in the following sections illustrate that there is a rich variety of useful designs that meet the condition. Hence, its theoretical shortcomings do not preclude practical usability.

In this chapter, it is assumed that no transitions contain multi-assignments, i.e., that the left-hand side of all assignments consists of a single state variable. Consider an arbitrary transition:

$$\mathit{TRANSITION}\quad t \quad \ll c \;\text{->}\; v := e \gg$$

From t the stability protocol, $stable_t(pre, post)$, is defined as follows:

$$stable_t(pre, post) \quad \equiv \quad (c.pre \wedge v.pre \neq e.pre) \Rightarrow \forall x \in R^t : same(x)$$

where $same(x)$ is a shorthand for $x.pre = x.post$, and R^t is the read set of the transition (defined in section A.14.1). Intuitively, $stable_t$ defines the constraint that none of the variables read by t change while t is active. This protocol is used to formulate the implementation condition stability.

> **Definition 7.4** *A design, D, meets the implementation condition stability, if and only if the following implication holds for all pairs of transitions,* t_1, t_2*, in D:*
>
> $$t_1(pre, post) \Rightarrow stable_{t_2}(pre, post)$$

When a design meets the stability condition, it is ensured that no active transition is affected by the state changes of other transitions.

Let D be a design that meets the stability condition. Then the Muller model (as defined in section 7.3) is speed-independent (according to definition 7.3). To see this, assume the contrary i.e., that there is a conflict state, s with two excited variables z_i, z_j such that

$$z_i = f_i(z_1, z_2, \ldots, z_j', \ldots, z_n) \text{ or } z_j = f_j(z_1, z_2, \ldots, z_i', \ldots, z_n)$$

Hence, there must exist two transitions, t_i, t_j,

$$\ll c_i \text{ -> } z_i := e_i \gg \qquad \ll c_j \text{ -> } z_j := e_j \gg$$

such that in the conflict state s

$$c_i.s \wedge z_i.s \neq e_i.s \bigwedge c_j.s \wedge z_j.s \neq e_j.s$$

It is assumed that

$$z_i = f_i(z_1, z_2, \ldots, z_j', \ldots, z_n) \text{ or } z_j = f_j(z_1, z_2, \ldots, z_i', \ldots, z_n)$$

But then there must be a state, s', such that

$$t_i(s, s') \wedge z_j.s' = e_j.s \quad \text{or} \quad t_j(s, s') \wedge z_i.s' = e_i.s$$

From this it follows that either $stable_i(pre, post)$ or $stable_j(pre, post)$ is not satisfied, and the stability condition is violated. This is a contradiction with the assumption that D meets the stability condition, and consequently, it can be concluded that D meets the definition of speed-independence.

Hence, for any design meeting the stability condition, the underlying circuit in the Muller model is speed-independent, and the term "speed-independent design" is used. Such a design allows for a speed-independent circuit realization.

Example: Verifying Stability

Consider the two simple designs shown in figure 7.5. These two designs are very similar and both describe an oscillator. However, the first ($oscc$) is speed-independent, and the second ($oscor$) is not. Consider first the cell $oscc$, where the stability protocols for the three transitions are

$$\begin{aligned}
stable_1(pre, post) &\equiv (a.pre = b.pre) \wedge (c.pre \neq a.pre) \Rightarrow same(a) \wedge same(b) \\
stable_2(pre, post) &\equiv (NOT\ c.pre \neq b.pre) \Rightarrow same(c) \\
stable_3(pre, post) &\equiv (NOT\ c.pre \neq a.pre) \Rightarrow same(c)
\end{aligned}$$

```
CELL oscc
   STATE a, b, c: BOOLEAN
BEGIN
   << a=b -> c:= a >> || (* t1 *)
   << b:= NOT c >> || (* t2 *)
   << a:= NOT c >> (* t3 *)
END oscc


CELL oscor
   STATE a, b, c: BOOLEAN
BEGIN
   << c:= a OR b >> || (* t4 *)
   << b:= NOT c >> || (* t5 *)
   << a:= NOT c >> (* t6 *)
END oscor
```

Figure 7.5 Two oscillators

To verify that the implementation condition stability is met, three cases must be considered, one for each of the three transitions t_i, $1 \leq i \leq 3$:

- Assume that $t_1(pre, post)$ holds. Then

  $$a.pre = b.pre \wedge c.post = a.pre \wedge same(a) \wedge same(b)$$

  It must be proven that $Stable_2(pre, post)$ and that $Stable_3(pre, post)$ hold. The first follows directly from considering the two cases $b.pre \neq NOTc.pre$ and $b.pre = NOTc.pre$. The second is proven by considering the two cases $a.pre \neq NOTc.pre$ and $a.pre = NOTc.pre$.

- Assume that $t_2(pre, post)$ holds. Then

  $$(b.post = NOTc.pre) \wedge same(a) \wedge same(c)$$

  It must be proven that $Stable_1(pre, post)$ and that $Stable_3(pre, post)$ hold. These follow directly from considering the two cases $b.pre \neq NOTc.pre$ and $b.pre = NOTc.pre$.

- Assume that $t_3(pre, post)$ holds. It must be proven that $Stable_1(pre, post)$ and $Stable_2(pre, post)$ hold. This is symmetric to the second case.

It follows that the first design ($oscc$) meets the implementation condition stability, and hence that it allows a speed-independent realization. This is not the case for the second design ($oscor$). Consider, for example, the first transition t_4, and assume that

$$c.post = (a.pre \vee b.pre) \wedge same(a) \wedge same(c)$$

In this case, $Stable_5(pre, post)$ does not hold, to see this consider, for example, a state where $a.pre = TRUE$, $b.pre = FALSE$, and $c.pre = FALSE$. It can, therefore, be concluded that the stability condition is not met and consequently, it cannot be expected that a circuit realization of the design $oscor$ is speed-independent.

7.3.2 Relationship with CREW

This section shows that the stability condition is stronger than the implementation condition CREW introduced in section 5.4.1. Both implementation conditions define a restricted class of design descriptions that allows a realization where transitions are executed simultaneously. However, in a self-timed circuit there is no external clock to ensure that signals are well defined; this must be ensured by the circuit itself. Therefore, to implement a design as a self-timed circuit, it is necessary to make more assumptions (restrictions) than to implement it synchronously.

Consider a design, D, that meets the stability condition. To show that D meets CREW, pick an arbitrary minimal cycle, c, in the read/write graph of D:

$$c = \{t_1, t_2, \ldots, t_n\}$$

To show that D meets CREW, it must be proven that

$$\neg(active(t_1) \wedge active(t_2) \ldots active(t_n))$$

where $active(t_i) \equiv c_i \wedge (v_i \neq e_i)$ (see section 5.4.1). None of the assignments of D is a multi-assignment (assumption on page 165); hence, $v_i \neq e_i$ implies that v_i is changed when t_i is executed. There are two cases to consider: either $v_i = e_i$ for all $1 \leq i \leq n$; in which case CREW is met or, if there is an i $(1 \leq i \leq n)$ such that $v_i \neq e_i$, then it follows from the stability condition that $NOT\, c_{i+1} \vee (v_{i+1} = e_{i+1})$. But then

$$\neg(active(t_1) \wedge active(t_2) \ldots active(t_n))$$

hence D meets CREW.

```
CELL oscc
  STATE a, b, c: BOOLEAN
  PROTOCOL
    a.pre=b.pre AND c.pre≠a.pre => same(a) AND same(b) AND
    NOT c.pre ≠ b.pre => same(c) AND
    NOT c.pre ≠ a.pre => same(c)
BEGIN
  ≪ a=b -> c:= a ≫ ||
  ≪ b:= NOT c ≫ ||
  ≪ a:= NOT c ≫
END oscc
```

Figure 7.6 Oscillator with stability protocol

It has now been shown that, stability is a stronger (more restrictive) implementation condition than CREW.

7.3.3 Mechanical Verification

This section describes how the mechanical theorem prover LP is used to check that a design is speed-independent. The implementation condition stability is introduced in section 7.3.1, and it is argued that it is a sufficient condition for speed-independence. The formulation of the stability condition makes it straightforward to check it mechanically. Such a mechanical is done in the same way as verifying invariants and protocols, because stability is formulated as a protocol. Consider the modification of the cell *oscc* (from section 7.3.1) shown in figure 7.6. The only difference is that a protocol is added. This protocol contains the conjunction of the three stability protocols, and it is called the combined stability protocol. It can be verified directly using the tools introduced in section 3.3, and for the simple example shown in figure 7.6 no further interaction with the theorem prover is needed. However, in more complicated cases it might be necessary to add additional protocols and invariants in order to verify that stability is met.

The stability protocol is a simple syntactical transformation of the transitions in a design, and it can be generated automatically from the design description in Synchronized Transitions. There is, however, a subtle difference between this protocol and the implementation condition stability. The combined stability protocol is the conjunction of *all* stability protocol (one for each transition). Hence, to verify the combined stability protocol, it is also necessary to prove that

$$t(pre, post) \Rightarrow stable_t(pre, post)$$

This excludes transitions that both read and write a state variable, for example, $\ll x := NOT x \gg$. However, such self-modifying transitions can be excluded without loss of generality. In [63] (section 5.3, theorem 5.1 and corollary 5.1), it is proven that if the circuit is speed-independent, then it can be designed without self-modifying transitions. Hence, there is no loss of generality in using the combined stability protocol to verify that a design is speed-independent.

7.3.4 Self-timed Signalling

In a self-timed circuit there is no global clock to determine when input is ready and when outputs are supplied. Instead, it must be possible to detect the arrival of a new input from the data themselves. To be able to do this, one uses a signalling convention where extra values (spacers), called empty values, are inserted between every pair of succeeding proper data values. Proper values are collectively called valid values.

To distinguish between empty and valid values, data are encoded. It is essential that the encoding of data ensures that no partially valid value is interpreted as a valid value; otherwise a computation can start prematurely. There are many ways of encoding data for self-timed computations [65]; most practical designs represent data in a dual-rail code where a pair of bits, $x.t, x.f$, is used to represent a single boolean x. The empty value (E) is represented by both bits low. The valid values true (T) and false (F) are represented by $x.t$ high and $x.f$ low and by $x.f$ high and $x.t$ low, respectively.

	$x.t$	$x.f$
E	low	low
F	low	high
T	high	low

Example: A FIFO Queue

Consider again a FIFO queue (see also section 6.2.2). Figure 7.7 shows a description of a self-timed FIFO. This design uses the self-timed signalling convention described above, so interface variables alternate between valid and

```
CELL fifo(STATIC n: INTEGER)
   TYPE
      vector = ARRAY [1..2*n] OF data
      range = [1..2*(n-1)]
   STATE q: vector
BEGIN
   {|| i: range | ≪ empty(q[i]) ≠ empty(q[i+2]) -> q[i+1]:= q[i] ≫ }
END fifo
```

Figure 7.7 A speed-independent FIFO

empty values, which is different from the other FIFO queues discussed in this book.

The type *data* includes the designated empty value, which is different from all valid values that the FIFO may hold. A boolean function *empty* is used, it returns *TRUE* if the argument is empty, and *FALSE* otherwise.

FUNCTION $empty(x : data)$: *BOOLEAN* *RETURN* $x=E$

The queue consists of a number of identical stages. Each of these alternates between holding a valid and an empty value. This alternation makes it possible to detect changes in a stage. In a stable configuration, the stages at the tail of the queue all hold the same value (empty or valid), and the nodes at the head alternate between empty and valid values as illustrated below:

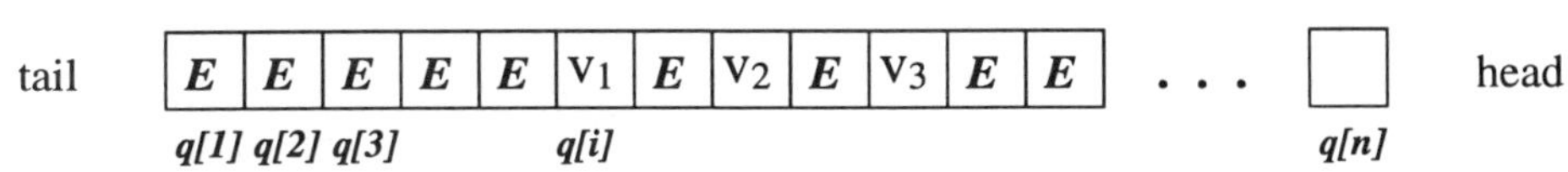

Input of a valid value can occur only when both $q[1]$ and $q[2]$ are empty. Similarly, input of an empty value can occur only when the first two stages hold a valid value. When a new valid datum is input to the queue, it is propagated to the right until it reaches a stage whose successor's output is not empty. In the above example, a new, valid input would propagate to stage $q[i-2]$; it cannot continue to $q[i-1]$, because $q[i]$ is valid. Likewise, an empty input value propagates to the right until it reaches a stage whose successor's output is empty.

Verifying that the FIFO is Speed-independent

The FIFO shown in figure 7.7 is speed-independent. This is verified by checking the implementation condition stability, which in this case is the following protocol:

PROTOCOL (stability *)*
$\{AND\ i:\ range\ |$
$(empty(q[i].pre) \neq empty(q[i+2].pre)\ AND\ (q[i+1].pre \neq q[i].pre))$
$\Rightarrow\ same(q[i])\ AND\ same(q[i+2])\}$

This protocol is satisfied by the FIFO design, but to verify it mechanically an additional invariant is needed, for example:

INVARIANT
$FORALL\ i:\ range\ .\ \ ((q[i] = q[i+1]) \Leftrightarrow (empty(q[i]) = empty(q[i+1])))$

The invariant excludes the possibility of having two different valid values next to each other in the queue. They must be separated by at least one empty value. This ensures that when $q[i+1]$ changes, the values of $q[i]$ and $q[i+2]$ remain unchanged, which is exactly what is required by the stability condition. If the invariant, stated above, is added to the design description shown in figure 7.7, then the stability protocol can be verified with ST2LP and LP.

7.3.5 A Refined FIFO

The design description of the FIFO queue given above specifies that every stage needs access to the values of both of its neighbors ($q[i]$ and $q[i+2]$). However, it is not necessary to transfer the complete value of the successor. All that is needed is to indicate whether the value is empty or valid, which can be indicated with a boolean. Such an indication is usually called an acknowledgement. Below, a refined FIFO is shown in which all stages are connected by forward-going data signals and a backward-going acknowledge signal.

TYPE link = RECORD $d:\ data\ \ a:$ *BOOLEAN END*

The refined FIFO design using the type *link* is shown in figure 7.8.

7.3.6 The C-element

The FIFO shown in figure 7.7 is based on a number of transitions of the form

```
CELL fifo(STATIC n: INTEGER)
  TYPE
    vector = ARRAY [1..2*n] OF link
    range = [1..2*n-1]
  STATE q: vector
  CELL stage(in, out: link)
  BEGIN
    ≪ empty(in.d)≠out.a -> out.d:= in.d ≫ ||
    ≪ in.a:= empty(out.d) ≫
  END stage
BEGIN
  {|| i: range | stage(q[i], q[i+1]) }
END fifo
```

Figure 7.8 A speed-independent FIFO

$$\ll empty(x) \neq empty(y) -> z := x \gg$$

Such a transition is frequently found in self-timed designs. It compares the values in x and y; when one is valid and the other is empty, the value of z is updated. Note that the order in which x and y change does not affect the computation. This is the key to the self-timed operation of the queue. There are a few variations of the transition that maintain this property:

$$\ll empty(x) = empty(y) -> z := x \gg$$
$$\ll x \neq y -> z := x \gg$$
$$\ll x = y -> z := x \gg$$

In fact, the oscillator (figure 7.1) shown as the very first design in this chapter uses such a transition. If x and y are of type boolean, then the last, of the three transitions shown above, describes a Muller C-element [54].

$$\ll x = y -> z := x \gg$$

This is a fundamental gate found in most self-timed circuits. It can be realized directly at the transistor level and is of approximately the same complexity as an AND gate. Sometimes, Muller C-elements with larger fan-in are used, for example, a three-input Muller C-element, which is described as follows:

$$\ll (x = y) \ AND \ (y = w) -> z := x \gg$$

```
CELL toggle(t, x, y: BOOLEAN)
  STATE z: BOOLEAN
  INVARIANT NOT (x AND y)
BEGIN
  ≪ (z≠y) AND (z=t) -> x:= z ≫ ||
  ≪ (z=x) AND (z≠t) -> y:= NOT z ≫ ||
  ≪ x≠y -> z:= y ≫
END toggle
```

Figure 7.9 A simple toggle

Example: The Toggle

The toggle considered in this example has an interface consisting of three variables t, x, and y. The variable t is interpreted as the input. Every other time t is true, x becomes true; otherwise y becomes true. After x or y becomes true, t may change to false again. When t is false, both x and y must become false. Then t can become true again, causing x or y to become true, etc.

Figure 2.17 (in the exercises for chapter 2) shows a representative sample of a computation. A toggle element can be built by combing three Muller C-elements as shown in figure 7.9. The state variable z records whether x or y changed the last time t changed. It is quite simple to verify that this toggle meets the stability condition, and hence that the design is speed-independent (see exercise 7.7).

The toggle element illustrates how a relatively weak invariant can be useful. It only states that x and y are never true simultaneously, but it does not constrain them to become true alternatively. Hence, the invariant does not express the intended functionality of the toggle (see exercise 7.8). Nevertheless, the invariant was very useful during the design; many faulty designs were rejected because the verification revealed that they did not satisfy the simple invariant. This stresses a point made several times in this book, namely, that incomplete but simple verification techniques can be a tremendous help in the design process.

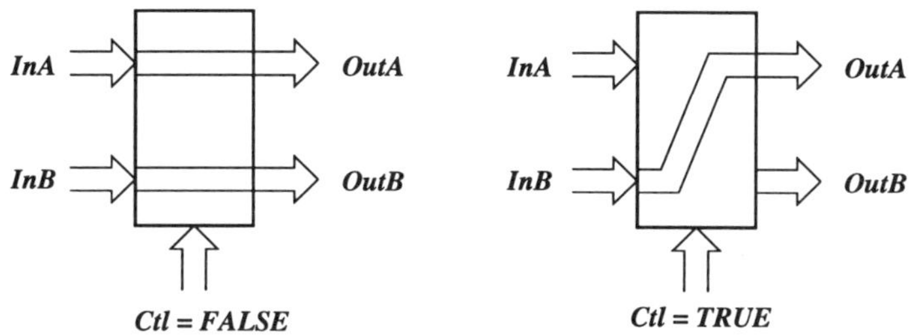

Figure 7.10 Two configurations of the switch

Example: The Switch

This section describes the design of a switch that multiplexes two data streams.[2] The data streams are input on two links inA, inB and output on the links $outA, outB$. A control link *ctl* determines the switching. When *ctl* is *FALSE*, the value of inA is passed directly to $outA$ and inB directly to $outB$. When *ctl* is $TRUE$, the value of inB is passed to $outA$ and the value of inA is ignored. The switch is illustrated in figure 7.10. It uses the same empty/valid protocol and links as the FIFO design shown in figure 7.8.

The switch shown in figure 7.11 consists of three parts: one for propagating data from the input to the output links, one for generating the acknowledge signals on the input links, and one for generating the acknowledge signal on the control link. It was quite tricky to make the design shown in figure 7.11. Several attempts failed before finding a design that is both functionally correct and speed-independent. Although the design description is only of modest size, it is too complex to be handled manually. In fact, an earlier version was designed and checked by hand and then sent for fabrication. Afterwards, a formal analysis revealed a problem in that design; this was corrected and led to the version shown in figure 7.11.

It is possible to mechanically verify that the switch design meets the stability condition. But to verify the the stability condition, it is necessary to find an invariant similar to what was the case with the FIFO design. However, the behavior of the switch is much more complicated, and so is the invariant. Quite a bit of effort was needed to come up with this auxiliary invariant. It is the result of an iterative process where an initial guess of what might be an invariant is gradually refined with the assistance of the theorem prover. When an attempt to verify a proposed invariant fails, the theorem prover can help in

[2] The design was first done by Jens Sparsø

```
TYPE
   elink = RECORD  d: data   a: BOOLEAN  END
   blink = RECORD  d: dual   a: BOOLEAN  END
CELL switch(Ctl: blink; InA, InB, OutA, OutB: elink)
STATE
 a0, a1: data
 aa1, ab1: BOOLEAN
BEGIN
 (* Data *)
   ≪ (Ctl.d=T) = empty(InB.d) –> a1:= InB.d ≫ ||
   ≪ (Ctl.d=F) = empty(InA.d) –> a0:= InA.d ≫ ||
   ≪ (Ctl.d=F) = empty(InB.d) –> OutB.d:= InB.d ≫ ||
   ≪ OutA.d:= IF empty(a0) THEN a1 ELSE a0 ≫ ||
 (* Acknowledgement of input data signals *)
   ≪ InB.a:= aa1 OR ab1 ≫ ||
   ≪ ((Ctl.d=T) = OutA.a) –> aa1:= OutA.a ≫ ||
   ≪ ((Ctl.d=F) = OutB.a) –> ab1:= OutB.a ≫ ||
   ≪ ((Ctl.d=F) = OutA.a) –> InA.a:= OutA.a ≫ ||
 (* Acknowledgement of the control signal *)
   ≪ ((Ctl.d=F) = InA.a) AND ((Ctl.d=F) = InB.a) –> Ctl.a:= InB.a ≫ ||
   ≪ ((Ctl.d=T) = InB.a) –> Ctl.a:= InB.a ≫
END switch
```

Figure 7.11 The switch design

strengthening the invariant by pointing out a counterexample to the proposed invariant.

In the switch the wave of values may spread over more than two elements, and the waves for valid and empty values are not the same. This is what makes the design of the switch tricky, so it should not be surprising that it also complicates the invariants. When the invariant is added to the design, it is quite simple to use ST2LP and LP to check that the switch meets the stability condition, and hence, that the design is speed-independent.

7.4 HIERARCHICAL DESIGNS

All the examples of speed-independent designs shown in the previous sections were single cells. More complicated examples lead to designs with a hierarchy of cells. This section shows how the localized verification technique described in section 3.4.1 is used to verify that such hierarchical designs are speed-independent.

To verify that the design is speed-independent, it must be verified that the design meets the stability protocol. This protocol is derived from the transitions of a cell and expresses constraints on both local state variables and formal parameters. As an example, consider the toggle design shown in figure 7.9. One of the conjuncts appearing in the stability protocol for this design is the following (stemming from the third transition):

$$((x.pre \neq y.pre)\ AND\ (z.pre \neq y.pre)) => same(x)\ AND\ same(y)$$

This is a local constraint on how x, y, and z may change in the cell. It is, however, also a constraint on the formal parameters, x, y, and hence, restricts the environment. The localized verification technique introduced in section 3.4 is also useful for verifying the stability protocols. The soundness of the localized technique ensures that all transitions throughout the design are consistent with all protocols (and invariants) including the stability protocol.

Example: The Modulo-n Counter

This section illustrates how speed-independence is verified using the localized verification technique. This is done by verifying the design of a modulo-n counter; this example was suggested by Ebergen [19]. The modulo-n counter has one boolean input, a, and two boolean outputs, p and q. Initially, all three are false. When the input a changes from false to true, the counter changes either p or q to true. The first $n-1$ times that a changes, p becomes true, but the n'th time it is q. When p or q are true, the environment must set a to false, and the counter must then ensure that both p and q become false. When all three are false, the environment may again set a true, and a new cycle starts. Figure 7.12 shows a speed-independent design of a modulo-n counter, where it is assumed that n is a power of two.[3] It is not difficult to design a counter that works for arbitrary n (see problem 7.10), but the general case takes more space. The simplified design shown in figure 7.12 illustrates the main point of this section, that speed-independence can be verified using localized verification. To save space, the invariants and protocols are not shown. Local invariance shows that all the transitions of the three cells (*toggle*, *count_even*, and *count_one*) maintain invariants and protocols, including the stability protocol. The non-interference step shows that none of the cell instantiations creates cells that violate invariants and protocols in other cells. Note that the verification of the stability protocol is treated as the verification of any other protocol.

7.5 DELAY-INSENSITIVITY

This sections discusses design of delay-insensitive circuits. As described in the introduction to this chapter, correct operation of a delay-insensitive circuit must not depend on assumptions about delays in wires connecting circuit elements. This is in contrast to speed-independent designs, where wires are assumed to have no delays.

First, a formal characterization of delay-insensitivity is given (this stems from [59]). The delay-insensitive circuits are a subset of the speed-independent circuits, so a more restrictive definition is needed. As mentioned in section 1.2, the computational model of Synchronized Transitions captures gate delays directly. Below it is shown how wire delays can be taken into account.

7.5.1 Modeling Wire Delays

When a state variable in a Synchronized Transitions design is modified, the new value is immediately available to all transitions that read the variable, because a variable name denotes the same value throughout the design description. However, the changes of a physical signal between logical values may not

[3] This design was done by Christian D. Nielsen

```
CELL n_count(a, p, q: BOOLEAN; STATIC n: INTEGER)
  RESTRICTION p, q: INTERNAL  a: EXTERNAL
  CELL toggle(t, x, y: BOOLEAN)
    STATE z: BOOLEAN
    INITIALLY z = TRUE
  BEGIN
    ≪ (z=t) AND (z≠y) -> x:= z ≫ ||
    ≪ (z=x) AND (z≠t) -> y:= NOT z ≫ ||
    ≪ x≠y -> z:= y ≫
  END toggle
  CELL count_even(a, p, q: BOOLEAN; STATIC n: INTEGER)
    STATE am, pm, qm, qm0, qm1, pm0: BOOLEAN
    INITIALLY p = FALSE  q = FALSE
  BEGIN
    ≪ am:= NOT (p OR q) ≫ ||
    ≪ a = pm0 -> p:= a ≫ ||
    ≪ a = qm1 -> q:= a ≫ ||
    ≪ pm0:= pm OR qm0 ≫ ||
    toggle(qm, qm0, qm1) ||
    n_count(am, pm, qm, n DIV 2)
  END count_even;
  CELL count_one(a, p, q: BOOLEAN)
    INITIALLY p=FALSE  q=FALSE
  BEGIN ≪ q:= a ≫ END count_one
BEGIN (* cell n_count *)
  {n = 1 | count_one(a, p, q) } ||
  {n>1 | count_even(a, p, q, n) }
END N_count
```

Figure 7.12 Design of a modulo-n counter

```
≪ x≠y' -> z:= y' ≫ ||
≪ x:= z ≫ ||
≪ y:= NOT z ≫ ||
≪ y':= y ≫
```

Figure 7.13 An oscillator with wire delay

occur at the same time for all circuit elements. There can be many reasons for this, for example, the time for a change of voltage to be transmitted along a wire, or variations in threshold voltages that cause elements to switch at different points of the change. To model this in the design description, it is therefore necessary to make a modification reflecting that changes to a state variable are not immediately available elsewhere in the design. This section shows one way of constructing such a model.

Let x be a state variable of a design. To model an arbitrary wire delay, a new state variable, x', is introduced, and the following wire transition is added to the design:

$$\ll x' := x \gg$$

Furthermore, one of the transitions that read x is modified by substituting x' for all occurrences of x. The nondeterminism in the execution of transitions models an arbitrary delay between assigning a new value to x and the detection of the new value. By introducing a new wire transition and state variable for each transition that reads x, changes of x can arrive at these transitions in any order. Consider again the transitions of the simple oscillator shown in figure 7.1.

$$\ll x \neq y \text{ -> } z := y \gg \ ||$$
$$\ll x := z \gg \ ||$$
$$\ll y := NOT z \gg$$

To model a delay on the wires carrying y, a wire transition is added and y' is substituted for y in the first transition (see figure 7.13). This design describes computations where the inversion of y can be delayed arbitrarily before propagating it to the transition comparing x and y'. If the behavior of the design shown in figure 7.13 is the same as that of the design in figure 7.1, then it can be concluded that the design is delay-insensitive with respect to one of the wires carrying the value of y. By introducing wire transitions for z and x, it is possible to analyze whether the design is also delay-insensitive with respect to the wires carrying these variables.

Example: A Delay-sensitive Design

Now consider the speed-independent design introduced in figure 7.4. Section 7.1 contains an informal argument that this design is not delay-insensitive. A more rigorous justification can be given by considering a design where two wire transitions are added:

$$\ll x \neq y' \;\text{->}\; z := y' \gg \;||$$
$$\ll y' \neq z \;\text{->}\; x := z \gg \;||$$
$$\ll z \neq x \;\text{->}\; y := x \gg \;||$$
$$\ll y' := y \gg \;||\; \ll y'' := y' \gg \;||$$
$$\ll (x = y'') \; AND \; (y'' = z) \;\text{->}\; stop := TRUE \gg$$

If the original design (see figure 7.4) is delay-insensitive then it performs the same computations as the one shown above. It is not difficult to verify that this is not the case. Assume that the design shown above starts in a state where

$$(x, y, z, y', y'') = (FALSE, TRUE, FALSE, TRUE, TRUE)$$

By performing the sequence of transitions

1. $\ll x \neq y' \;\text{->}\; z := y' \gg$
2. $\ll z \neq x \;\text{->}\; y := x \gg$
3. $\ll y' := y \gg$
4. $\ll y' \neq z \;\text{->}\; x := z \gg$

the design enters a state where

$$(x, y, z, y', y'') = (TRUE, FALSE, TRUE, FALSE, TRUE)$$

In this state, the last transition is enabled, and performing it makes $stop$ true. This computation cannot occur in the original design. It can therefore be concluded that the computations of the two designs are not the same, and hence this is not a delay-insensitive design.

This example shows how added wire transitions are useful for verifying that a design is *not* delay-insensitive. Below, the same transformation is used to formally define delay-insensitivity.

7.5.2 Definition of Delay-insensitivity

In the preceding sections it is shown how to model delays in wires and gates. Based on this model, it is now possible to give rigorous definition of delay-insensitivity. Consider the transformation of a design D to a design D' with one wire transition. Let S be the set of state variables of D, T be the transitions of D, x be a variable in S, and t be a transition in T that reads x. The design with the x input of t arbitrarily delayed, D', has state variables S' and transitions T' such that

$$\begin{array}{rcl} T' & = & T - \{t\} \cup \{t'\} \cup \{\ll x' := x \gg\} \\ S' & = & S \times x' \end{array}$$

where t' is obtained from t by replacing all occurrences of x with x'. If D' can be obtained from D by a transformation of this form, then D' is a single-delayed version of D. D_n is a delayed version of D_1 if there exists a finite sequence of designs, $D_1, \ldots D_n$, such that for each $i : 1 < i \leq n$, D_i is a single-delayed version of D_{i-1}. note that this definition allows cascaded wire transitions for the same state variable.

Definition 7.5 *A design D is delay-insensitive if and only if every delayed version of D is a refinement of D.*

The above definition of delay-insensitive is called the "foam-rubber principle" [52], because it requires the circuit to behave correctly under arbitrary extensions of wires (foam rubber). As illustrated by the example given in section 7.5.1, the addition of a wire transition can be used to verify that a design is *not* delay-insensitive. It is more difficult to do the reverse, i.e., verifying that a design is delay-insensitive. In principle, an arbitrary number of wire transitions must be considered; this is a cumbersome if not impossible approach, because an arbitrary number of delayed designs must be considered. It is possible to formulate implementation conditions that are sufficient for ensuring delay-insensitivity. As was the case with the other implementation conditions, CREW and stability, the objective is to give the designer the possibility of checking the needed restrictions on a high-level design. A detailed presentation of the implementation conditions for delay-insensitivity is beyond the scope of this book. In [59] the conditions are introduced and used to check that the FIFO presented in section 7.3.4 is delay-insensitive.

Example: A Delay-insensitive Oscillator

In simple cases, it is possible to use definition 7.5 to verify that a design is delay-insensitive. This is illustrated using the oscillator shown in figure 7.1. There are three state variables in this design, and one of these (z) is read by two different transitions; hence the possible wire delays can be modeled by introducing four wire transitions with corresponding new state variables (zz, zzz, xx, yy). This yields the delayed version of the design shown in figure 7.14. It is

```
CELL oscillator
   STATE x, y, z, zz, zzz, xx, yy: BOOLEAN
   INITIALLY x=TRUE  y=FALSE  z=TRUE
BEGIN
   ≪ xx≠yy  ->  z:= yy ≫ ||
   ≪ zz:= z ≫ || ≪ zzz:= z ≫ ||
   ≪ x:= zz ≫ || ≪ xx:= x ≫ ||
   ≪ y:= NOT zzz ≫ || ≪ yy:=y ≫
END oscillator
```

Figure 7.14 Delayed version of oscillator

quite simple to verify that this design is a refinement of the original oscillator. The verification can be done mechanically using the translator mentioned in chapter 6 and LP. To do the verification mechanically, it is necessary to add the following invariant:

INVARIANT

$$((x = z) \Rightarrow (z = zz))\ AND\ ((y = (NOT z)) \Rightarrow (z = zzz))\ AND$$
$$((z \neq x) \Rightarrow (x = xx))\ AND\ ((z = y) \Rightarrow (y = yy))$$

For simple designs, this procedure can be used to formally verify that a design is delay-insensitive; however, for more complicated designs it is not practical.

7.6 BACKGROUND

Many of the ideas of asynchronous circuits were first developed in the 1950s. The Muller model used in section 7.3 stems from this period [54]. It was the basis for many practical and theoretical developments. The primary concern was gate delays, and therefore the emphasis was on speed-independent circuits. At that time, it was quite common to use asynchronous techniques. An example is the Atlas computer [38] built in Manchester.

Recently, there has been a revived interest in self-timing, partly for technological reasons and partly because the properties of self-timed circuits have inspired many theoretical computer scientist. With today's VLSI technology, wire delays can be significant either because the wires are long or because of other technological effects, such as differences in switching thresholds. Consequently, there is more emphasis on delay-insensitive circuits in the recent research. Most of the theoretical work is based on the foam-rubber principle presented in section 7.5 [52, 59, 62].

The emphasis in the theoretical research has been on characterizing and checking speed-independence and delay-insensitivity. Typically, this is done using a model checker [17, 39], which is a significant difference from the approach taken here based on a theorem prover. There are also a number of proposals for synthesizing speed-independent or delay-insensitive circuits from high-level descriptions. There are two main directions in this research. One is based on high-level programming like languages such as CSP [18, 47] or Synchronized Transitions [12]. The other is based on Petri-net-like graphs, for example, Signal Transition Graphs [13].

There are also a number of recent experimental results. In his recent Turing lecture [61], Sutherland presented "transition signalling" as an approach for building self-timed pipelines and gave many examples. His designs depend on matching the delays in data paths to delays in the corresponding control circuitry. At Caltech, Alain Martin and his colleagues have explored another approach to designing delay-insensitive circuits, which like Synchronized Transitions is based on a concurrent model. They have obtained impressive results, including the design of a simple self-timed microprocessor [48]. Their approach is based on a specification in a high-level language using message passing and applying *systematic transformations* to obtain the final design.

The implementation condition stability presented in section 7.3 has some similarity to a condition given by Chandy and Misra [10] who used the term delay-insensitive (which is not consistent with the terminology used here).

7.7 EXERCISES

Exercise 7.1 Consider the following design:

CELL $oscillator$
  STATE x, y, z : *BOOLEAN*
  INITIALLY $x = FALSE \quad y = TRUE \quad z = FALSE$
BEGIN
  $\ll x \neq y \texttt{->} z := y \gg \|$
  $\ll y \neq z \texttt{->} x := z \gg \|$
  $\ll z \neq x \texttt{->} y := x \gg$
END $oscillator$

Verify that this design maintains the invariant $NOT((x = y)\ AND\ (y = z))$.

Exercise 7.2 Consider the following transitions:

$\ll x \neq y \texttt{->} x := 2 * y \gg$
$\ll NOT\,reql\ AND\ grl \texttt{->} grl, grr, reqp := FALSE, reqr, reqr \gg$
$\ll from \neq empty\ AND\ to = empty \texttt{->} to, from := from, empty \gg$

Give the stability protocol for each of the transitions.

Exercise 7.3 A half-adder is a combinational circuit with two input data bits a, b and two output data bits s, c, for which

$s = a \neq b \qquad c = a\ AND\ b$

Describe a speed-independent design of a half-adder where a, b, c, and s are encoded using dual-rail coding. Verify that the design is speed-independent.

Exercise 7.4 A data-type $vector$ is defined as follows:

STATIC $n = ...$
TYPE
  $range = [1..n]$
  $vector$ = *ARRAY* $range$ *OF BOOLEAN*

Design a cell ccc that is a n-input Muller C-element:

CELL $ccc(a : vector; y :$ *BOOLEAN; STATIC* n : *INTEGER)*

When all n elements of a have the same value, then this is propagated to y; otherwise y is not changed.

Exercise 7.5 Design a simplified version of the switch shown in figure 7.11 where one input stream *in* is multiplexed between two output streams *outT*, *outF* determined by the value of a control signal *ctl*. When *ctl* is true, the output stream *outT* is used; when *ctl* is false, *outF* is used.

Exercise 7.6 Find a design that meets the implementation condition CREW but not stability.

Exercise 7.7 What is the stability protocol for the design shown in figure 7.9? Use this protocol to verify that the toggle element is speed-independent.

Exercise 7.8 One of the requirements to the toggle shown in figure 7.9 is that x and y alternate. Can this be expressed as an invariant? Can it be expressed as a protocol?

Exercise 7.9 Assume that the type *data* used in a link (see section 7.3.5) is a single dual-rail value, e.g.,

TYPE
  dual = *RECORD* t, f : *BOOLEAN END*
  element = *dual*

Describe the function *empty* in Synchronized Transitions. Give a realization of a single FIFO stage using only Muller C-elements and standard gates (such as NOT, OR, etc.).

Exercise 7.10 Design a cell, *oddcount*, similar to the cell *evencount* shown in figure 7.12, which works for an odd value of n. Use the cell *oddcount* to generalize the design shown in figure 7.12 to a modulo-n counter that works for an arbitrary n.

Exercise 7.11 Use refinement to verify that the toggle shown in figure 7.9 is delay-insensitive.

8

TOWARDS LARGER DESIGNS

The previous chapters show how to use Synchronized Transitions for a variety of design tasks: modelling, verification, refinement, etc. The emphasis has been on introducing and motivating the language as well as the underlying computational model. Therefore, all examples have been relatively small focusing on a single aspect at a time. Practical examples are usually larger and involve a mixture of concerns. This chapter discusses two problems which one may encounter in a larger design: one is mixing synchronous and asynchronous computations, the other is combining software and hardware realization of different components of a single design.

8.1 COMBINING ASYNCHRONOUS AND SYNCHRONOUS COMPUTATIONS

This book contains many examples of both synchronous and asynchronous computations. However, all examples given in the other chapters are either synchronous or asynchronous. In fact, the language report in appendix A explicitly states that it is not allowed to mix the synchronous and asynchronous combinators. This restriction simplifies the language, the computational model, and the verification techniques. However, it also limits the practical usability of Synchronized Transitions which is illustrated by the two examples that follow.

```
TYPE buffer = RECORD s: BOOLEAN; d: element END
CELL producer(out: buffer)
  STATE
    t: INTEGER
    next: element
  INITIALLY t = 0
BEGIN
  ≪ t:= (t+1) MOD 10 ≫ +
  ≪ t ≤ 9 -> next:= P(t, next) ≫ +
  ≪ t = 9 -> out.s, out.d:= TRUE, next ≫
END producer
CELL consumer(in: buffer)
  STATE
    s: INTEGER
    l: element
  INITIALLY t = 0
BEGIN
  ≪ s:= (s+1) MOD 10 ≫ +
  ≪ s > 0 -> l:= C(t, l) ≫ +
  ≪ s = 0 -> in.s, l:= FALSE, in.d ≫
END consumer
```

Figure 8.1 Producer and consumer cells

Example: Producer/Consumer

Consider a design consisting of two synchronous processes, called a producer and a consumer. The producer supply data which is sent to the consumer for further processing. The following design description shows how data is produced:

$$\ll t := (t+1)\ MOD\ 10 \gg\ +$$
$$\ll t \leq 9 -> next := P(t, next) \gg$$

P is a function (not specified in detail here) which gradually transforms the data item contained in the variable $next$. After ten iterations, the producer has finished its processing, and the value of $next$ is handed over to the consumer, which also needs ten iterations to complete its processing. This processing is done by the function C (which is not specified here). Figure 8.1 shows a design of two cells, one for the producer and one for the consumer. If the producer and consumer are composed synchronously, they will then operate in lock step. This models a situation where both are realized as synchronous circuits controlled by the same global clock. Now consider a realization where the producer and consumer are controlled by different clocks, for example, a realization where they are located on different circuit boards connected to a common bus or a realization, where they are located far from each other and connected by a communication network. To model such realizations, the producer and consumer are connected by an asynchronous FIFO (similar to the one described in section 6.2.2).

```
CELL fifo(in, out : buffer)
  STATE q : ARRAY [0..n-1] OF buffer
  TRANSITION transfer(to, from : buffer)
    << from.s AND NOTto.s -> from.s, to := FALSE, from >>
BEGIN
   transfer(in, q[0])  ||  transfer(q[n-1], out) ||
   {|| i : [0..n-2] | transfer(q[i], q[i+1]) }
END fifo
```

The complete design consisting of the producer, the consumer, and the FIFO is described as follows:

$$producer(p) \;\; || \;\; fifo(p, c) \;\; || \;\; consumer(c)$$

However, in this design description synchronous and asynchronous composition is mixed which is not allowed according to the language report in appendix A. However, neither a purely synchronous nor a purely asynchronous computation are adequate models. The synchronous is too restrictive, because the producer and the consumer are forced to operate in lock step, while in the asynchronous model, it is not possible to control the number of steps used for producing a data item. Syntactically, it seems straightforward to allow designs where the combinators are mixed, and judging from the FIFO example, it also seems simple to construct a computational model.

A possible computational model can be illustrated with the following generic example:

$$
\begin{array}{l}
(t_{11} + t_{12} + \ldots\ t_{1n_1}) \,||\\
(t_{21} + t_{22} + \ldots\ t_{2n_k}) \,||\\
\quad\ldots\\
(t_{k1} + t_{k2} + \ldots\ t_{kn_k})
\end{array}
$$

This describes a transition system with k groups of transitions which describes a computation with repeated nondeterministic selection of a synchronous group of transitions, e.g., $t_{i1} + t_{i2} + \ldots + t_{in_i}$, followed by a synchronous execution of all the transitions in the selected group. This is a generalization of the asynchronous model which is repeated nondeterministic selection and execution of a single enabled transition. The only difference is that groups of transitions and not single transitions are executed. This model is useful for analyzing the producer/consumer example. It could, for example, be used to reveal synchronization errors or other violations of safety properties; but it cannot be used to analyze qualitative aspects like buffer capacity. The next section present another example of a mixed synchronous/asynchronous design.

Although the producer/consumer example itself is quite small, it illustrates a problem which is often encountered in large designs. The producer and consumer are separate synchronous computations which can be realized by synchronous circuitry controlled by different clocks. Such separated synchronous computations are forced to communicate asynchronously, for example, using an asynchronous FIFO. These globally asynchronous, locally synchronous realizations are a hybrid using both self-timed circuitry and synchronous circuitry. This compromise can take advantage of some of the benefits of self-timing without the overhead associated with speed-independent or delay-insensitive circuitry. In [31] is shown a nontrivial example of such a design, which was developed using Synchronized Transitions.

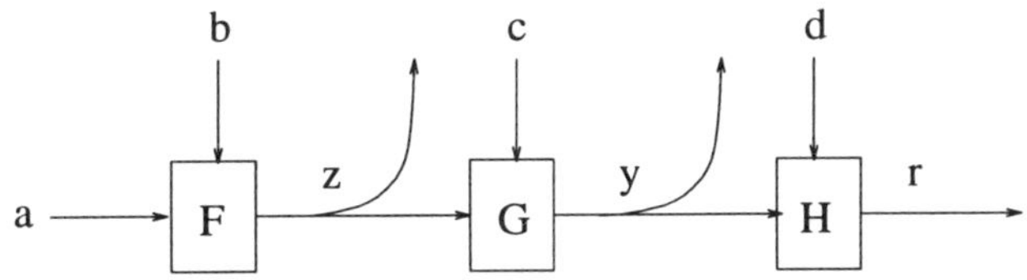

Figure 8.2 A simple pipeline

Example: A Simple Pipeline

This section shows an example of how it can be useful to mix asynchronous and synchronous computations. Consider the simple pipelined computation shown in figure 8.2.

$$\ll z := F(a, b) \gg \;\; || \;\; \ll y := G(z, c) \gg \;\; || \;\; \ll r := H(y, d) \gg$$

Functions such as F, G, and H model combinational subcircuits, and the asynchronous composition models a circuit where the composition is also combinational. Now assume that the pipeline is placed in a synchronous environment, for example one, where the output r is fed back to the input a through a synchronous latch, and where the number of iterations are counted, e.g.,

$$\ll a{:}= r \gg \; + \; \ll i{:}= i{+}1 \gg$$

In every iteration, the value of r is fed back into a, and i is incremented. Figure 8.3 shows a design description where the two parts are combined. The intended interpretation is an iterative computation where each iteration consists of evaluating the combinational part ($r := H(G(F(a, b), c), d)$) yielding a new value of r, updating a (once), and incrementing i. It is quite similar to a synchronous computation, each part is executed once in every iteration. Executing the combinational part consists of repeating the three asynchronously composed transitions until no further state changes are possible (a local fixed point).

As was the case with the producer/consumer it is not allowed to mix syn-

```
(≪ z:= F(a, b) ≫ || ≪ y:= G(z, c) ≫ || ≪ r:= H(y, d) ≫)
  +
≪ a:= r ≫ + ≪ i:= i+1 ≫
```

Figure 8.3 Counting iterations of a pipeline

chronous and asynchronous parts. Although the intuitive meaning may seem obvious in these examples, it is not so simple to find a general computational model which encompass both the examples.

Computational Model of Mixed Designs

The two previous examples have illustrated the usefulness of mixing the asynchronous and synchronous combinators in a design description. In the two simple cases, it was not difficult to give a computational model of the mixed designs, however, in the general case, it is not so straightforward to establish a computational model which is:

- simple enough to be of practical use,
- powerful enough to capture a number of nontrivial examples, and
- useful for formal analysis and verification.

Several approaches have been tried, see for example [31]. However a thorough treatment of this topic is beyond the scope of the present book. The next section describes another aspect of making large designs.

8.2 CODESIGN

Today, an advanced electronic product typically consists of a mixture of hardware and software components. Therefore, the functionality of the product must be partitioned into components realized in different technologies and these components must be interfaced. However, hardware and software are typically designed separately, using different design techniques, different design tools, and by different people. Figure 8.4 a) illustrates this approach, where a product is designed by separating the hardware and software development at a very early phase of the design. But making the partitioning early in the design often makes it difficult to decide what should be realized in hardware and what should be realized in software. Putting too much in hardware may increase the design costs and putting too much in software may create performance problems. In practise, it is often just an educated guess based on past experience. If it was possible to make an abstract model of a design covering both hardware and software, a codesign, one could instead do as suggested in

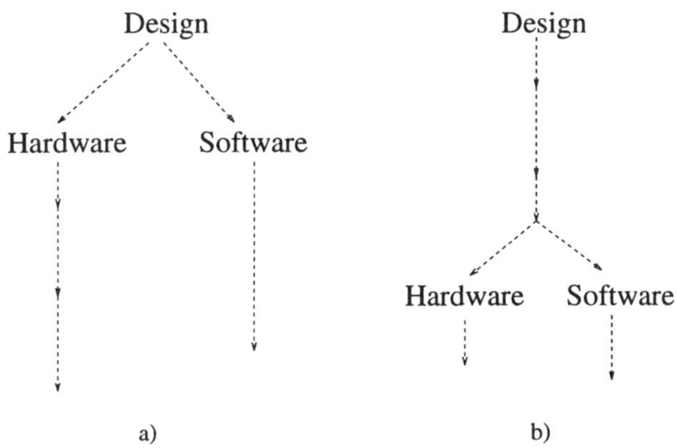

Figure 8.4 Early binding of hardware and software versus late binding

figure 8.4 b), where a large part of the development is done without binding the components to a particular kind of realization.

An abstract model of a codesign must be general enough to cover both hardware and software components, and it must enable a designer to describe his design without enforcing a commitment as to what should be realized as software and what should become hardware. The abstract models provided by design descriptions in Synchronized Transitions are very similar to the models used for software design. It is, therefore, tempting to try using such models for codesign also, and the next section gives an example of a computation described in Synchronized Transitions which can be realized in both hardware and software. To simplify the discussion the generic terms hardware and software are used to cover a whole range of possible realizations such as, general purpose processors, signal processors, FPGAs, or full-custom VLSI circuits. There are a number of other alternatives which makes it difficult to draw a sharp line between hardware and software.

Example: A Priority Queue

This section describes the design of a priority queue. Even though the example is quite simple, it illustrates the advantages of having an abstract design description which can be realized as either hardware, software, or a combination of the two.

A priority queue is a data structure that holds a set of elements from a domain with an ordering relation, <, hence, any two elements can be compared. There are two external operations on a priority queue: insertion and removal. New elements can be inserted at any time (unless the queue is full). Removal takes out the *smallest* of the elements currently in the queue. Hence, the design of the priority queue must ensure that when an element is removed, it is indeed the smallest (according to the ordering of the elements).

The interface to the priority queue consists of two registers: an input register, *in*, and an output register, *out*. Each of these registers may contain an element from the domain of the priority queue, and each has additional status information indicating whether the register is full or empty. An element may be inserted in the queue when the input register is empty. Similarly, an element may be removed when the output register is not empty.

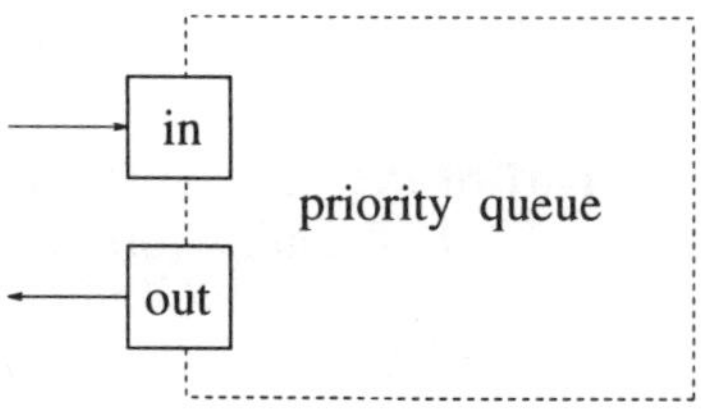

Priority queues have many applications, for example, in event driven simulators where the priority queue stores all elements scheduled for execution. The priority is in this case the starting time of the event to be simulated. The event driven simulation algorithm always executes the event with the smallest (earliest) starting time, which is supplied by the priority queue. Priority queues are also used in several efficient graph algorithms, e.g., graph traversal. In an event driven simulator, the operations on a priority queue are often the bottleneck of the algorithm, hence, it is important to use an efficient realization of the priority queue. In software, it can be implemented so that the execution time of insertions and removals is proportional to the logarithm of the number of elements stored in the priority queue. However, even with such a realization, the priority queue can be the bottleneck, when the event queues are large and the operations frequent, as is the case in circuit simulators. It can, therefore, be advantageous to realize the priority queue in hardware, which makes the simulator a codesign where most of the simulation algorithm is realized in software, while the time critical priority queue is realized in hardware. This partitioning is discussed in further detail below, but first the behavior of the priority queue is described using Synchronized Transitions.

The interface to the priority queue consists of the two registers (*in*, *out*). A real-

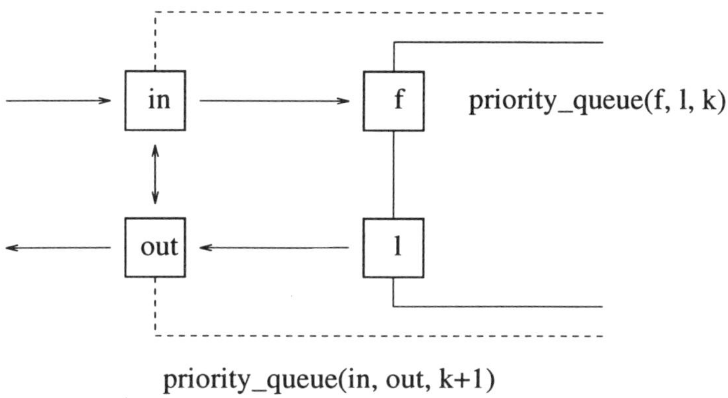

Figure 8.5 Extended priority queue

ization of a priority queue has a limited capacity defined by a static parameter, d.

CELL $priority\_queue(in, out:\ element;$ *STATIC* d : *INTEGER)*

The type element is a record with two fields, one of type boolean, s, representing the status of the register ($TRUE$ if the register is full and $FALSE$ if it is empty). The other field, p, represents the priority of the element (low numbers are removed first) and is of type integer.

TYPE $element$ = *RECORD* p : *INTEGER* s : *BOOLEAN END*

Assume that $priority\_queue(f, l, k)$ is a priority queue; this queue can be used to construct a priority queue with a larger capacity by extending it with two additional registers, in and out (see figure 8.5). The behavior of the extended priority queue is described as follows:

$\ll in.p < out.p \rightarrow in, out := out, in \gg \ ||$
$\ll NOT f.s\ AND\ (in.p \geq out.p) \rightarrow in.s, f := FALSE, in \gg \ ||$
$\ll NOT out.s\ AND\ (in.p \leq l.p)\ AND\ (in.p \leq f.p) \rightarrow in.s, out := FALSE, in \gg \ ||$
$\ll NOT out.s\ AND\ (l.p \leq f.p)\ AND\ (l.p \leq in.p) \rightarrow l.s, out := FALSE, l \gg$

The first transition puts a newly inserted element (contained in the variable in) in front of the queue if it is smaller than out (and thereby the smallest element in the entire queue). Alternatively, the second transition inserts the element stored in the variable in into the priority queue $priority\_queue(f, l, k)$, thus making room for a new insertion ($in.s = FALSE$). When an element is removed from the queue, out becomes empty, and the third or fourth transition places in out the element that is now the smallest element of the priority queue. This is either the newly inserted element in in, or the smallest element of the

```
CELL priority_queue(in, out: element; STATIC d: INTEGER)
   STATE f, l: element
   INITIALLY f.s=FALSE l.s=FALSE
BEGIN
   ≪ in.p < out.p –> in, out:= out, in ≫ ||
   ≪ NOT f.s AND (in.p ≥ out.p) –> in.s, f:= FALSE, in ≫ ||
   ≪ NOT out.s AND (in.p ≤l.p) AND (in.p ≤f.p) –> in.s, out:= FALSE, in ≫ ||
   ≪ NOT out.s AND (l.p ≤f.p) AND (l.p ≤in.p) –> l.s, out:= FALSE, l ≫ ||
   {d > 0 | priority_queue(f, l, d-1) } ||
   {d = 0 | ≪ f.p < l.p –> f,l:= l, f ≫ ||
            ≪ NOT l.s AND f.s –> f.s,l:= FALSE, f ≫ }
END priority_queue
```

Figure 8.6 Priority queue design

priority queue $priority\_queue(f, l, k)$. The net result is a priority queue with capacity $k + 1$ using a priority queue with capacity k. A recursive description of a priority queue following this outline is shown in figure 8.6.

The invariant shown below captures the key property of the priority queue, namely that the element stored in *out* is smaller than all elements in the priority queue $priority\_queue(f, l, d)$ represented by the variables f and l.

INVARIANT $out.p \leq l.p\ AND\ out.p \leq f.p$

In [60] it is shown how to verify this invariant using ST2LP and LP.

The priority queue specified above can be realized in hardware as well as in software. A hardware realization would have two externally visible registers: in, out, and a number of internal registers corresponding to $f, l, \ldots$. Furthermore, a number of comparators are needed, see figure 8.7. It is also simple to realize the priority queue in software. Such a realization is not presented here, but it can be found in many introductory texts on data structures and algorithms (for example, [14]). For a large priority queue, the hardware realization gives the fastest operations, however, it is also the most expensive both in terms of development costs and in terms of circuitry. It is cheaper to make a software realization where the elements are stored in standard off-the-shelve memory components. The traffic is most intense at the head of the queue. Towards the end, there is only traffic when the queue is almost full. It could, therefore, be advantageous to realize the first registers in hardware and use software to

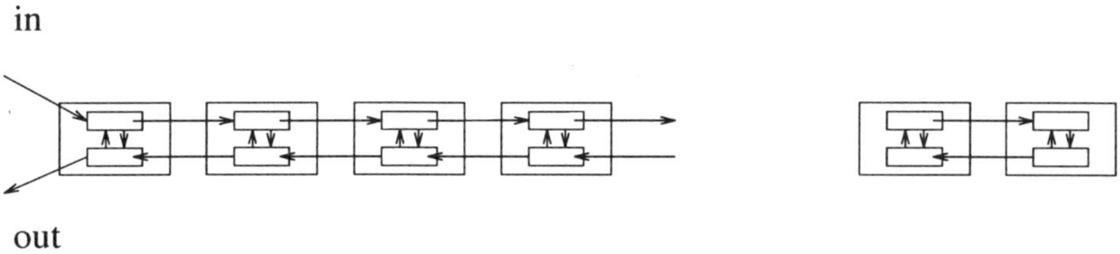

Figure 8.7 Hardware realization of a priority queue

realize the rest. The exact partitioning depends on the application and the speed required.

For the priority queue, a combination of software and hardware might be the optimal realization. By giving an abstract description of the computation without commitments to a particular kind of realization, it becomes possible to map the description to hardware, software, or combinations of the two. The last possibility requires an efficient interface between the hardware and the software, and this issue is discussed in section 8.2.2. Before addressing this, it is described how to realize a Synchronized Transitions design in software.

8.2.1 Software Realizations

A design description in Synchronized Transitions is quite similar to a program in a high-level programming language, and it is not difficult to transform a design description into an executable program. Such transformations are useful both for realizing the software parts of a codesign and for simulating a hardware design described in Synchronized Transitions. This section illustrates the translation into Pascal, and it is straightforward to make a similar translation to other high-level languages.

A design description consists of state variables and transitions. The declarations of state variables are translated directly into variable declarations of Pascal. Consider the following generic transition:

TRANSITION t_i $\ll c_i \text{->} v_i := e_i \gg$

The state change performed when executing such a transition is easily expressed in Pascal:

PROCEDURE t_i
BEGIN *IF* c_i *THEN* $v_i := e_i$ *END*

The computational model of asynchronous designs assumes that transitions are executed one at a time, but no specific order is required. Hence, any order is acceptable. A design description with n transitions can, therefore, be executed by a loop where one of the transitions is selected and executed in each iteration.

```
REPEAT
  select := random(1, n);
  CASE select OF
    1 : t1 ;
    2 : t2 ;
    ...
    i : ti ;
    ...
    n : tn
  END
FOREVER
```

The procedure *random* is a random number generator which is used for selecting a transition in each iteration. For synchronous designs, it is necessary to use a different realization ensuring that *all* transitions are executed in each iteration, and that all transitions use the pre-state for evaluating expressions. This can be realized by duplicating all state variables and temporarily storing the new values of a state variable v in its duplicate v'. The transition t_i shown above is realized with the following Pascal procedure:

```
PROCEDURE ti
BEGIN  IF ci THEN vi' := ei  END
```

The new value of the state variable v is assigned to v' which means that the old value of v is obtained in all expressions where it appears. The main loop of the simulation is also a bit different:

```
REPEAT
{INVARIANT : v1 = v1'  v2 = v2'  ... vk = vk' }
    t1 ;
    t2 ;
    ...
    ti ;
    ...
    tn
    v1 := v1'; v2 := v2'; ... ; vk := vk';
FOREVER
```

The last line of the loop updates all state variables with their new values. If the invariant:

$$v_1 = v'_1 \quad v_2 = v'_2 \quad \ldots v_k = v'_k$$

is maintained, it is not necessary to discern between the state variables that were actually changed in the last iteration and those that were not changed.

The software realizations shown above are used in translator called ST2C which transforms a design description into a C program. This is used for simulating designs written in Synchronized Transitions. To simulate large design with cells and quantifications, one may either do all the instantiations and create a flat design or use the instance tree (these alternatives were also considered for verification in chapter 3). In ST2C the first alternative (flattening) is used, but for large designs this is not practical.

The software realization is a simple way of generating an executable program from a design description in Synchronized Transitions; however, the resulting program is sometimes inefficient because of the sequencing of transitions. Efficient algorithms often rely on a clever ordering of operations. The all pairs shortest paths problem discussed in section 2.6 is a good example. If the transitions are executed in the order indicated by the nested loop shown below then all pairs of shortest paths are computed in n^3 steps:

```
FOR i := 1 TO n DO
  FOR j := 1 TO n DO
    FOR k := 1 TO n DO
      IF d[i][k] + d[k][j] < d[i][j] THEN d[i][j] := d[i][k] + d[k][j]
```

This is Warshall's algorithm [66] which is found in many introductory textbooks on efficient algorithms. It illustrates that the sequencing of operations can be very important for the efficiency of a program.

8.2.2 Interfaces

This section discusses how to model and realize the interface between hardware and software components. It is assumed that the partitioning into hardware and software is done at the level of cells, i.e., that some cells become hardware and others become software. This is by no means the only possibility, others are working on partitioning at the level of a single transition [3]. Although this question of granularity is interesting, it is not addressed here. Instead, the emphasis is on the interface between hardware and software, because it is

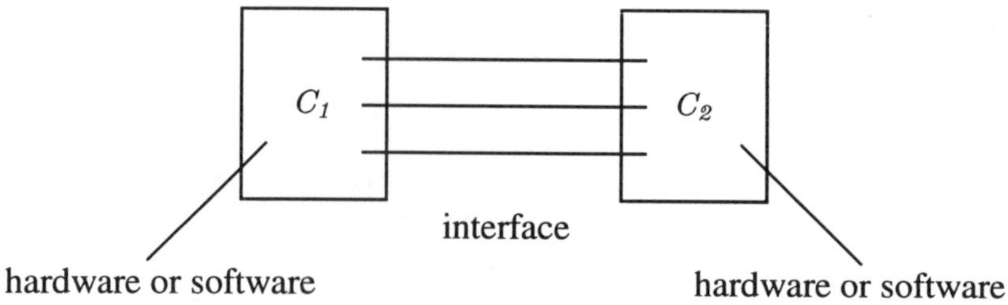

Figure 8.8 Interface between two components

important that the abstract model of the interface can be mapped efficiently to both hardware and software. If this is not the case, it is easy to loose the performance advantage obtained by realizing some components as special purpose hardware. Figure 8.8 shows two components and their interface, each is a cell which can be realized either as hardware or as software. The interface of a cell is its parameter list which becomes a set of state variables when the cell is instantiated. The realization of the state variables is different in hardware and software, but the interpretation in the abstract model is the same. This possibility of a dual realization is an important attribute of state variables when used to interface hardware and software. In software, state variables are realized as program variables, i.e., represented in memory, and in hardware they are realized by latches. In either case, there is no difference between the realization of an interface variable and all other (internal) variables which means that there is no additional overhead associated with a state variable used to communicate between hardware and software. In software, this is quite similar to handling input/output by memory mapping where certain areas in the address space have special properties, for example, that an i/o-driver can access it.

The tools (compiling to software or synthesizing circuitry) must of course be aware of which variables that make up the interface and allocate them accordingly, for example, by memory mapping where some locations in the address space can also be referenced by the synthesized hardware.

8.3 BACKGROUND

The combination of synchronous and asynchronous components are hard to avoid, it is for example present whenever a synchronous circuit interfaces with exterior unclocked systems such as a human operator. There are some fundamental limitations in such computations [46], and there are some difficult design problems to resolve. It is, therefore, both important and also difficult to find adequate models of such mixed computations.

Codesign is currently a very active research topic [57] and among the many issues which are under investigation is the choice of design language. To do codesign, one needs a model and a corresponding language which is abstract enough to allow descriptions of both hardware and software. However, almost all existing models and languages aim at either describing software (programming languages) or circuits (hardware description languages). Both kinds of languages aim at describing a *computation*, but they are usually based on very different concepts. Almost all programming languages are based on the Von-Neumann model, where a computation is described by a *sequence* of instructions operating on data stored in a memory. An instruction counter keeps track of which instruction in the sequence to execute next. However, in most circuits there are many parts operating in parallel; so sequencing is at best an extra obstacle, and in some cases it might even preclude an efficient realization. For a hardware designer, the spatial relationship between different components is a primary concern; therefore, many hardware description languages emphasize structural descriptions where the physical organization of parts is explicitly specified. As it was pointed out in the introduction, the intention with Synchronized Transitions is to abstract from both sequencing and structure when describing a computation.

9

EPILOG

This book describes a computational model that has fascinated me, since I first encountered it, almost ten years ago. I have gradually learned that despite its apparent simplicity, this concurrent transition model is a powerful conceptual tool with a broad range of applications as illustrated by the hardware oriented examples shown in this book and by the software oriented examples given in Chandy and Misra's book on UNITY [10]. Furthermore, the model opens for a range of interesting manipulations, for example, formal verification, timing analysis, and synthesis. Each of these can be done in other models also, but using the concurrent transition model, a range manipulations are available using a *single* model.

The concurrent transition model is the most important aspect of the present book, and Synchronized Transitions is a language for describing such models in a succinct formal notation. The key constructs of the language are transitions, quantification, and cells. These constructs are all rather different from what is available in more traditional programming and hardware description languages. Another important difference is the lack of a sequencing operator. By excluding this, the formal manipulations are simplified, because a design can be handled by considering the transitions one at a time. This is illustrated both by the verification and the synthesis techniques. To complete Synchronized Transitions, additional constructs are needed, for example, types, expressions, and declarations. This book only uses a very simple type concept which is a restricted version of what is available in other languages. It is very likely that a more useful design language can be obtained by combining the concurrent transition model with more powerful constructs for data modeling.

The concurrent transition model can express computations that are difficult to realize as efficient circuits. It is, for example, not always easy to implement the atomicity. Rather than restrict the model or complicate the realization, the concept of implementation conditions plays a key role. In this book two classes of circuits have been discussed in some detail, synchronous and speed-independent realizations. For each of these an implementation condition is defined which makes it possible to check whether it is feasible to realize a particular design before proceeding with a detailed realization. For both implementation conditions it is shown how to mechanize the checking of the condition.

The collection of examples is another important aspect of the book. The significance of the examples goes beyond their use as illustrations; they also form a small "literature" that shows how to express various commonly occurring structures in Synchronized Transitions. Most of the examples are at the circuit level, e.g., an adder and an arbiter, which can be explained and presented using only a few pages. More substantial and higher level examples would be more convincing, but they would also require more space, both to explain the problems and their solutions. By publishing this book, I hope that others, like me, will become fascinated by the concurrent transition model and try it on more more substantive applications, thus creating a richer literature.

The book also describes some secondary and more technical contributions, in particular the pre-/post-state formalization which is the basis for all the verification techniques. First and foremost, the localized technique on which the mechanization is based. The pre-/post-state formalization has turned out to be very fruitful. I first used it for mechanizing the verification of invariants. Out of this grew the idea of protocols, the refinement conditions, and the technique to verify speed-independence.

Although most of my effort during several years have been devoted to developing the approach described here, there are still many and significant limitations. The lack of large high-level examples has already been mentioned. From a theoretical point of view, I think that the most serious limitation is that synchronous and asynchronous designs cannot be mixed. Many others will consider the emphasis on safety properties to be a serious shortcoming. One may well ask: "what is the point of verifying some safety properties, if it cannot be ensured that the final realization will perform even a single step of the specified computation?" Two answers are offered: First, there are many important and desirable properties that are not verified formally, for example, power consumption, timing, and noise sensitivity. Presently, liveness of designs in Synchronized Transitions is among these. Secondly, a safety proof ensures that if a realization ever delivers an output, then it meets the verified properties. This is a significant guarantee. In practice, many design errors manifest themselves as circuits giving the wrong answer rather than as not giving any answer at all.

It would of course be nice, if one could find more complete verification techniques, possibly also some covering liveness aspects. However, if the completeness is obtained at the expense of the simplicity or computational complexity, the price might be too high. In practice there are many other factors setting the limits of what can be handled formally. Given the choice between completeness and simplicity, I have given highest priority to the latter.

A

SYNCHRONIZED TRANSITIONS REPORT

This appendix gives a detailed description of all constructs of the Synchronized Transitions language except expressions and types, which are not covered. The syntax of Synchronized Transitions is defined formally by a grammar (written in a variation of BNF), whereas the static semantics is defined informally in natural language. Nonterminals are written in lower case, e.g., *cell-declaration* and terminals are either written in upper case, e.g., *CELL*, or enclosed in quotes, e.g., ")". Constructs enclosed by square brackets are optional and if enclosed in braces they may be repeated zero or more times. A vertical bar (|) separates alternatives.

Any implementation of Synchronized Transitions has a rigid definition of types and expressions, which can be found in the documentation accompanying the particular tools. This documentation also specifies possible limitations of the tool.

A.1 DESIGN

A design in Synchronized Transitions is a cell declaration. This declaration can be preceded by static and type declarations.

design = {type-declarations [";"] | named-static-declarations [";"] }
cell-declaration [";"]

A.2 DECLARATIONS

The declarations define a number of named quantities.

declaration = cell-declaration |
type-declarations |
named-static-declarations |
state-declarations |
initially-part |
transition-declaration |
function-declaration |
invariant-declaration |
fixpoint-declaration |
protocol-declaration |
restriction-part

A.3 CELLS

A cell is the major compositional unit. It defines a name, an interface, and a block, possibly consisting of a number of (inner) cells.

cell-declaration = CELL name [interface] [";"]
block
END name
block = {declaration [";"] } BEGIN cell-body
cell-body = instantiations

The cell body defines the instantiations that take place when the cell itself is instantiated. Instantiation of a cell describes how locally declared state variables are created and performs the local instantiations of the cell (which may lead to instantiations of possible inner cells). A cell may instantiate itself recursively, but recursion is a description mechanism only. The recursion takes place during instantiation.

The name of the cell is visible in the block where the cell declaration appears (including inner cells). The local part of a cell is the block excluding all inner cell declarations. The name occurring immediately after *END* must be the name of the cell.

A.4 NAMED STATIC DECLARATIONS

A named static is an identifier denoting a value.

named-static-declarations = STATIC
named-static-declaration {[";"] named-static-declaration }
named-static-declaration = name "=" expression

The type of the named static is determined by the type of the expression which must be a static expression. The expression determines the value of the named static and this value cannot be changed. Named static declarations must not be recursive.

A.5 STATE VARIABLE DECLARATIONS

A state variable is a named data structure that holds a value which can change dynamically.

state-declarations = STATE
variable-declaration {[";"] variable-declaration }
variable-declaration = namelist ":" named-type

A state variable holds a value that can be changed by a transition. It may only hold values in the set associated with the type (of the variable). When the name of a state variable appears in an expression it denotes the last value assigned to it, called the current value.

A.6 INITIALIZATION

Locally declared state variables can be initialized.

initially-part = INITIALLY multi-initialization {[";"] multi-initialization }
multi-initialization = initial-list | quantified-initialization
quantified-initialization = "{" range-declaration "|" initial-list "}"
initial-list = left-hand-initial-list "=" right-hand-initial-list
left-hand-initial-list = designator-list
right-hand-initial-list = expression-list

The initially part defines the initial values of state variables. An initialized state variable holds that value until the first change is made by a transition. State variables are not required to be initialized. A quantified initialization defines a number of similar initializations. The range declaration specifies a range index and a set of values that it takes. For each of these values the initializations (after the |) are made. The range index is a static. The scope of the range index is the initializations only. If the name of the range index is already declared, then that declaration becomes invisible in the scope of the range index. The type of a range index is always integer. The range specified in the quantification only specifies which initializations are to be made; not the type of the range index.

Statics must not appear as designators. They may, however, appear as operands in expressions, for example as index expressions to array elements in the left-hand list. The number of elements in the left- and right-hand lists must be the same, and they must be pairwise type equivalent (pairing the first designator with the first expression, the second designator with the second expression, etc.). The expressions in the right-hand list must all be static.

A.7 TRANSITION

Transitions specify state changes.

transition-declaration = TRANSITION name [interface] [";"]
transition-instantiation
transition = "≪" transition-body "≫"
transition-body = [precondition "–>"] multi-assignment |
precondition
precondition = expression

The multi-assignment is performed only when the precondition is true; the transition is then said to be enabled. Transitions are executed atomically. A transition need not be executed immediately after its precondition becomes true, and there is no upper bound on when it occurs. When the precondition is omitted, it is by default statically true. A transition without a multi-assignment cannot in itself affect the computation. The type of the precondition must be type equivalent to boolean.

A.7.1 Multi-assignments

A multi-assignment specifies new current values for one or more state variables.

multi-assignment = left-hand-list ":=" right-hand-list
left-hand-list = designator-list
right-hand-list = expression-list

The multi-assignment is executed by first evaluating all expressions in the right-hand list, these values are then assigned to the designators in the left-hand list. The first designator gets the value of the first expression, the second designator gets the value of the second expression, etc.

All designators in the left-hand list must be distinct. Statics may not appear as designators, they may, however, appear as operands in expressions. This includes index expressions in the left-hand list in the case that the designator is an array element. The number of elements in the left- and right-hand lists must be the same, and they must be pairwise type equivalent (pairing the first designator with the first expression, the second designator with the second expression, etc.). If the right-hand side is a static then its type must be equivalent to the type of the designator on the left-hand side.

A.8 FUNCTION DECLARATION

A function declaration defines a parameterized procedure for computing a value.

function-declaration =
    FUNCTION name [interface] ":" named-type [";"] function-body
function-body = RETURN expression | protocol-expression

The function body defines an expression that is evaluated every time the function is called (appears in an expression). The value obtained by evaluating the expression is returned as the function value.

The value is considered to be of state kind, even if the expression in the function body is static. The type of the expression must be equivalent to the named type given in the function declaration.

If the body expression is an ordinary expression then the function is called an ordinary function. If the body expression is a protocol expression then the function is a called a protocol function. As some expressions can be used both as ordinary and protocol expressions, the same is the case for functions.

Recursive calls of the function in the function body are not allowed.

A.9 FORMAL PARAMETERS

A formal parameter list defines the interface to a parameterized construct (cell, transition, or function).

interface = "(" [formal-parameters] ")"
formal-parameters = formal-declaration {";" formal-declaration }
formal-declaration = [STATE | STATIC] namelist ":" named-type

There are two kinds of formal parameters: state parameters and static parameters; state is default. A state parameter is used as a state variable whereas a static parameter is used as a named static. A static parameter may have different actual values in different instantiations/calls.

A.10 INVARIANT DECLARATION

An invariant is a boolean expression that should be satisfied in all states of a computation.

invariant-declaration = INVARIANT expression

An invariant is used for verifying the design, it has no influence on the execution. Two or more invariants in the declaration part of the same cell is treated as one invariant being the conjunction of the invariants involved.

The type of the expression must be type equivalent to boolean.

A.11 PROTOCOL DECLARATION

A protocol is an expression that should be satisfied after each execution of a transition.

protocol-declaration = PROTOCOL protocol-expression

A protocol is used for verifying the design, it has no influence on the execution. Two or more protocols in the declaration part of the same cell is treated as one protocol, being the conjunction of the protocols involved. Protocol expressions can specify relations between the values of state variables before and after execution of a transition.

The type of the expression must be type equivalent to boolean.

A.12 FIXED POINT DECLARATION

A protocol is an expression that should be satisfied by any fixed point state of a design.

fixed-point-declaration = FIXPOINT expression

A fixed point is used for verifying the design, it has no influence on the execution. The type of the expression must be type equivalent to boolean.

A.13 RESTRICTION PART

Restrictions specify limitations on where a state variable can be changed.

restriction-part = RESTRICTION restriction {[";"] restriction }
restriction = namelist ":" restriction-kind
restriction-kind = EXTERNAL | LOCAL | INTERNAL | UNRESTRICTED

All names appearing in the namelist must be state variables or state parameters, and they must be visible by the scope rules. The four kinds of restrictions have the following meaning:

- external: the variable may only be changed externally, i.e., in a surrounding cell instance (this restriction is meaningful only for formal parameters),
- local: the variable may only be changed locally, i.e., in neither internal nor external cell instances,
- internal: the variable may only be changed in internal cell instances, i.e., by instances to which it is passed as an actual parameter,
- unrestricted: there are no restrictions on the variable; it may be changed anywhere.

If a state parameter or state variable is not mentioned in any restriction part, it is by default unrestricted. The following table shows the valid combinations of restrictions on actual parameters and formal parameters.

Actual parameters	Formal parameters
external	external
local	external
internal	external local internal unrestricted
unrestricted	external unrestricted

When an actual parameter is restricted to be internally written, it can be written in any cell instance to which it is passed as a parameter and in which the corresponding formal parameter is not restricted to be external. If the formal parameter in one of these instances is restricted to be local or internal, then the formal parameters in all other instances must be restricted to be external to ensure that the restrictions are not violated. That is, a formal locally or internally restricted parameter may not be instantiated with an actual parameter that can be changed in another instance in the outer cell.

A.14 INSTANTIATION

The instantiation part specifies how transitions and cells are created.

instantiations = instantiation {combinator instantiation }
instantiation = cell-instantiation |
transition-instantiation |
transition |
quantified-instantiation |
conditional-instantiation |
"(" instantiations ")"
cell-instantiation = name [actual-parameters]
transition-instantiation = name [actual-parameters]

When a cell is instantiated a fresh copy of all local state variables is created and the instantiations in the cell body are done. Actual parameters are bound to the formal parameters in the cell body. Instantiation of a transition creates a new instance of the transition body where the actual parameters are bound to formal parameters.

A quantified instantiation is used to create a number of similar instances.

quantified-instantiation =
"{" combinator range-declaration "|" instantiations "}"
range-declaration = name ":" range

The range declaration specifies a range index and a set of values that it takes. For each of these values the instantiations (after the |) are made. The range index is a static, so it may be passed to a cell or transition instantiation as a static parameter. The scope of the range index is the instantiations only. If the name of the range index is already declared, then that declaration becomes invisible in the scope of the range index. The type of a range index is always integer. The range specified in the quantification only specifies which instances are to be made, not the type of the range index.

A conditional instantiation is used to make instantiations depend on the value of a static expression.

conditional-instantiation = "{" expression "|" instantiations "}"

The expression must be static and yield a boolean result. The instantiations are made only if the result is true.

A.14.1 Combinators

The combinators are used for composing transitions and cell instantiations.

combinator = "||" | "*" | "+"

The asynchronous and synchronous combinators cannot be mixed in one design. The first, ||, called asynchronous composition specifies independent and atomic execution. The second, +, called, synchronous composition specifies that all enabled transitions are executed simultaneously. In a synchronous design all preconditions of all transitions are evaluated, then all assignments of the enabled transitions are performed simultaneously. Synchronous composition of a set of transitions is only allowed if no two transitions can assign a value to the same designator at the same time, as specified by the exclusive write condition below. Consider the n transitions: t_i where $0 < i \leq n$,

TRANSITION $t_i \ll c_i \;\text{->}\; v_i := e_i \gg$

The write set, W^i of the i'th transition is the set of designators appearing in v_i. Similarly, the read set, R^t, of the i'th transition is the set of designators appearing in c_i and e_i.

The exclusive write condition on the n transitions is defined as follows:

Exclusive Write Condition: The transitions $t_1, t_2, \ldots t_n$ meet the exclusive write condition if and only if:

$$\forall i_1, i_2 \in [1..n] : W^{i_1} \cap W^{i_2} \neq \emptyset \Rightarrow \neg(c_{i_1} \wedge c_{i_2})$$

To compose the transitions $t_1, t_2, \ldots, t_n$ synchronously they must satisfy the exclusive write condition. When cells are composed synchronously the transitions in these cells must satisfy the exclusive write condition.

The * combinator provides a mechanism for factoring either synchronously or asynchronously composed transitions. The product composition of two transitions meeting the exclusive write condition is defined as follows:

$$\ll c_1 \text{ -> } l_1 := e_1 \gg + \ll c_2 \text{ -> } l_2 := e_2 \gg \equiv \ \ll c_1 \wedge c_2 \text{ -> } l_1, l_2 := e_1, e_2 \gg$$

The following algebraic laws hold for the three combinators. In the following I_j denotes an arbitrary instantiation.

Commutative:	$I_1 \| I_2$	$\equiv$	$I_2 \| I_1$
	$I_1 * I_2$	$\equiv$	$I_2 * I_1$
	$I_1 + I_2$	$\equiv$	$I_2 + I_1$
Associative:	$(I_1 \| I_2) \| I_3$	$\equiv$	$I_1 \| (I_2 \| I_3)$
	$(I_1 + I_2) + I_3$	$\equiv$	$I_1 + (I_2 + I_3)$
	$(I_1 * I_2) * I_3$	$\equiv$	$I_1 * (I_2 * I_3)$
Precedence:	$I_1 \| I_2 * I_3$	$\equiv$	$I_1 \| (I_2 * I_3)$
	$I_1 + I_2 * I_3$	$\equiv$	$I_1 + (I_2 * I_3)$
Distributive:	$I_1 * (I_2 \| I_3)$	$\equiv$	$I_1 * I_2 \ \| \ I_1 * I_3$
	$I_1 * (I_2 + I_3)$	$\equiv$	$I_1 * I_2 \ + I_1 * I_3$

A.15 ACTUAL PARAMETERS

When a cell or transition is instantiated or a function is called, actual parameters are bound to formal parameters.

actual-parameters = "(" [actual-parameter {"," actual-parameter }] ")"
actual-parameter = designator | expression

When doing an instantiation or function call the actual values of the parameters are determined. Expressions (including those appearing in designators) are evaluated. The first actual parameter is bound to the first formal parameter, the second actual to the second formal, etc. Changing the value of a formal state parameter also changes the value of the corresponding actual parameter (and vice versa).

The actual parameters of the cell instances in the cell must obey the restriction rules. The number of actual parameters must be exactly the same as the number of formal parameters specified in the declaration. Furthermore, the actual and formal parameters must be pairwise type equivalent. An actual static parameter is a static expression. Statics may not appear as actual state parameters, they may, however, appear as operands in expressions. Conversely, designators may not appear as actual static parameters. In the case of a cell instance or transition, an actual state parameter must be a designator. In function calls actual state parameters may be either designators or state expressions.

A single state variable must not be known under two (or more) different names in the scope of a cell (no aliasing). Therefore, all actual state parameters passed to a cell instance must be distinct.

A.16 KINDS

There are two kinds of expressions: state and static. An expression that do not contain any designators and function calls is a static expression.

A state expression is an expression that is not a static expression. It may contain statics, function calls, and designators.

A.16.1 Quantified Expression

A quantified expression is a short notation for a repeated use of an operator on a sequence of operands.

quantified-expression = "{" ac-operator range-declaration "|" expression "}"
quantified-protocol-expression =
    "{" ac-operator range-declaration "|" protocol-expression "}"

An ac-operator is an associative and commutative dyadic operator. The quantified expression:

$\{@i : [m..n] \mid e(i)\ \}$

is the same as:

$e(m)@e(m+1)@...@e(n)$

If the lower limit, m, is larger than the upper limit, n, then the value of the quantified expression is the neutral element for the operator @. The order of evaluation the expressions ($e(i)$) is not specified. @ must be an ac-operator. If the expression (after the |) of a quantified expression is static then the quantified expression itself is static. The result type is type equivalent to the type of this expression.

A.17 PROTOCOL EXPRESSIONS

A protocol is a boolean expression that must be satisfied after each execution of a transition. The expression relates the values of state variables before and after execution of a transition.

protocol-expression =
protocol-expression dyadic-operator protocol-expression |
monadic-operator protocol-expression |
static |
designator "." (pre | post) |
function-call |
"(" protocol-expression ")" |
quantified-protocol-expression |
conditional-protocol-expression

A protocol expression relates the values of state variables before and after execution of a transition. The value of a state variable before a transition is executed is written as its designator followed by *.pre*, and its value afterwards is denoted by *.post* following the designator. The function call must be a call of a protocol function.

Statics can either be protocol expressions or ordinary expressions.

A.18 DESIGNATORS

A designator denotes a state variable either directly or indirectly through the parameter mechanism.

designator-list = designator { "," designator }
expression-list = expression { "," expression }
namelist = name { "," name }

A designator is a state variable or formal state parameter.

Two designators are distinct, if and only if:

- they have distinct names or
- they are distinct elements of a structured variable.

Two names are distinct if they are different and one of them is not an alias (renaming through parameter passing) of the other.

A.19 FUNCTION CALL

A function call yields a value by evaluating the expression of a function.

function-call = name "(" actual-parameters ")"

The name must be the name of a function. A function call may be an ordinary expression or a protocol expression. A function call is an expression of kind state. The type is equivalent to the result type in the function declaration.

A.20 STATICS

A static denotes a value that is fixed before execution starts.

static = name | literal

The name must denote a named static, formal static parameter, or range index.

A literal is a specification of a value.

literal = symbolic-boolean | symbolic-integer
symbolic-boolean = TRUE | FALSE
symbolic-integer = [sign] digit {digit }
digit = "0" | "1" | "2" | "3" | "4" | "5" | "6" | "7" | "8" | "9"

A static may appear as a static expression.

A.21 NAMES, BLOCKS, AND SCOPE RULES

Each named quantity has a scope which is the part of the design description where the definition of the name applies. A block is a part of the text introduced by declarations, quantified expressions, and instantiations. A block may contain inner blocks, which are considered part of the block. There are the following kinds of blocks:

- A cell block, which is the text between the parameter list and the *END* of the cell, but excluding the name of the cell.
- A transition block, which is the text between the parameter list and the terminating "≫" of the transition, but excluding the name of the transition.
- A function declaration block, which is the text between the parameter list and the end of the expression, excluding the name of the function.
- A quantified expression block, which is the quantified expression, excluding the interval of quantification.
- A quantified instantiation block, which is the quantified instantiation, excluding the interval of quantification.
- An quantified initialization block, which is the quantified initialization excluding the interval of quantification.

The local part of a block is the block excluding inner blocks. The names declared in the local part of a block are called local names, and these must all be distinct. A declaration is visible when the named quantity has the properties (type, value, etc.) specified by the declaration.

The scope of a name is contained in the block where the name is declared, furthermore it depends on what the name denotes:

state variable, function, and transition names: The scope is the block in which the declaration occurs, including (local) function and transition blocks, but excluding inner cell blocks. This means that variables declared in enclosing cells are not visible in a cell, unless explicitly passed as parameters. Functions and transitions are not visible in inner cells.

symbolic static, type, and cell names: The scope is the block in which the declaration occurs, including inner blocks. Therefore, a symbolic static, type or cell name declared in an enclosing cell is visible in a cell.

formal parameters (function or transition): The scope is the function or transition block.

formal static parameters (cell): The scope rules are the same as for locally declared symbolic statics.

formal state parameters (cell): The scope rules are the same as for locally declared state variables.

range index names: The scope is the instantiation of the quantified instantiation, quantified expression, or quantified initialization where the range index is declared. The interval of quantification is not part of the scope.

All names can be given a new definition in an inner block, in which case their previous declaration becomes invisible in the block of the new declaration. So, for instance, static, type, and cell names can be given new definitions in inner blocks where they would otherwise be visible. State variable, function, and transition names can be given a new definition as formal parameters in a (local) function or transition block, or as range indexes and components in inner quantification and record blocks. When a name is declared in two overlapping blocks, it is always the *innermost* declaration which applies.

REFERENCES

[1] R.F. Ayers. *VLSI Silicon Compilation and the Art of Automatic Chip Design*. Prentice-Hall, 1983.

[2] Ralph J.R. Back and R. Kurki-Suonio. Distributed cooperation with action systems. *ACM Transactions on Programming Languages and Systems*, 10(4):513–554, 1988.

[3] Edna Barros and Wolfgang Rosenstiel. A clustering approach to support hardware- / software partitioning. In Jerzy Rozenblit and Klaus Buchenrieder, editors, *Codesign: Computer-Aided Software/Hardware Engineering*. IEEE Press, 1993.

[4] Robert S. Boyer and J. Strother Moore. *A Computational Logic*. Academic Press, 1979.

[5] Robert S. Boyer and J. Strother Moore. *A Computational Logic Handbook*. Academic Press, 1988.

[6] R.K. Brayton, G.D. Hachtel, C.T. McMullen, and A. Sangiovanni-Vincentelli. *Logic Minimization Algorithms for VLSI Synthesis*. Kluwer Academic Publishers, 1984.

[7] Richard P. Brent and H. T. Kung. A regular layout for parallel adders. *IEEE Transactions on Computers*, 31(3):260–264, March 1982.

[8] Raul Camposano and Wayne Wolf. *High-level VLSI Synthesis*. Kluwer Academic Publishers, 1991.

[9] K. Mani Chandy. Concurrent programming for the masses. In *Proceedings of the 4th Annual ACM Symposium on Principles of Distributed Computing*, pages 1–12. ACM, 1985. Invited address presented at the 3rd Annual ACM Symposium on Principles of Distributed Computing.

[10] K. Mani Chandy and Jajadev Misra. *Parallel Program Design: A Foundation*. Addison-Wesley, 1988.

[11] Jacob Christensen. Efficient synthesis of hierarchical circuit descriptions. Master's thesis, Department of Computer Science, Technical University of Denmark, 1990.

[12] Per Henrik Christensen, Henrik Hulgaard, and Jørgen Staunstrup. Synthesis of delay insensitive circuits from verified programs. In J.-P. Banatre and D. Le Metayer, editors, *Research Directions in High-level Parallel Programming Languages*, pages 326–337. Springer Lecture Notes 574, 1992.

[13] T.A. Chu, C.K.C. Leung, and T.S. Wanuga, editors. *A Design Methodology for Concurrent VLSI Systems*. Proceedings of ICCD, 1985.

[14] Rick Decker. *Data Structures*. Prentice Hall, Inc., 1989.

[15] Peter Denyer and David Renshaw. *VLSI Signal Processing: A Bit-serial Approach*. Addison-Wesley, 1985.

[16] Edsger W. Dijkstra. Guarded commands, nondeterminacy, and formal derivation of programs. *Communications of the ACM*, 18(8):453–457, 1975.

[17] David L. Dill. Trace theory for automatic hierarchical verification of speed-independent circuits. In Jonathan Allen and F. Thomson Leighton, editors, *Advanced Research in VLSI: Proceedings of the Fifth MIT Conference*, 1988.

[18] Jo C. Ebergen. A formal approach to designing delay-insensitive circuits. *Distributed Computing*, 5(3):107–119, 1991.

[19] Jo C. Ebergen and Ad M. G. Peeters. Modulo-N counters: Design and analysis of delay-insensitive circuits. In Jørgen Staunstrup and Robin Sharp, editors, *2nd Workshop on Designing Correct Circuits, Lyngby*, pages 27–46. Elsevier, 1992.

[20] E.A. Emerson. Temporal and modal logic. In Jan van Leeuwen, editor, *Handbook of Theoretical Computer Science*, volume B, pages 995–1071. Elsevier, 1990.

[21] Richard Lipton et al. VLSI layout as programming. *ACM Transactions on Programming Languages and Systems*, 5(3):405–421, 1983.

[22] R.W. Floyd. Assigning meanings to programs. In J.T. Schwartz, editor, *Proceedings of the Symposium in Applied Mathematics*, volume 19, pages 19–32. American Mathematical Society, 1967.

[23] Michael P. Fourman. Formal system design. In Jørgen Staunstrup, editor, *Formal Methods for VLSI Design*, pages 191–236. North-Holland/Elsevier, 1990.

[24] Michael P. Fourman and Robert L. Harris. Lambda – logic and mathematics behind design automation. In *26th ACM/IEEE design Automation Conference*, 1989.

[25] Stephen J. Garland and John V. Guttag. An overview of LP: the Larch Prover. In *Proceedings of the Third International Conference on Rewriting Techniques and Applications*. Springer-Verlag, 1989.

[26] Stephen J. Garland and John V. Guttag. A guide to LP, the Larch Prover. Technical report, Digital SRC, 1991.

[27] Stephen J. Garland, John V. Guttag, and James J. Horning. Debugging Larch shared language specifications. *IEEE Transactions on Software Engineering*, September 1990.

[28] Stephen J. Garland, John V. Guttag, and Jørgen Staunstrup. Verification of VLSI circuits using LP. In *The Fusion of Hardware Design and Verification*, pages 329–346. IFIP WG. 10.2, North Holland, 1988.

[29] Mike Gordon. HOL A Proof Generating System for Higher-Order Logic. *VLSI Specificacion, Verification and Synthesis*, 1987.

[30] Mike J. C. Gordon and Tom F. Melham. *Introduction to HOL*. Cambridge University Press, 1993.

[31] Mark R. Greenstreet. Using Synchronized Transitions for simulation and timing verification. In Jørgen Staunstrup and Robin Sharp, editors, *Proceedings from DCC'92, IFIP Transactions A-5*, pages 215–236. Elsevier, 1992.

[32] Aarti Gupta. Formal hardware verification methods: A survey. *Formal Methods in Systems Design*, 1(2/3):151–238, October 1992.

[33] John V. Guttag, James J. Horning with S.J. Garland, K.D. Jones, A. Modet, and J.M. Wing. *Larch: Languages and Tools for Formal Specification*. Springer-Verlag Texts and Monographs in Computer Science, 1993. ISBN 0-387-94006-5, ISBN 3-540-94006-5.

[34] C.A.R. Hoare. Proof of correctness of data representations. *Acta Informatica*, 1(1):271–281, 1972.

[35] C.A.R. Hoare. Communicating sequential processes. *Communications of the ACM*, 21(8):666–677, August 1978.

[36] IEEE, New York. *VHDL Language Reference Manual*, std 1076-1987 edition, 1988.

[37] D. Johnson. Finding all the elementary cycles of a digraph. *SIAM Journal of Computing*, 4:77–84, 1975.

[38] T. Kilburn, D.B.G. Edwards, M.J. Lanigan, and F.H. Sumner. One-level storage system. *IRE Transactions on Electronic Computers*, EC-11:223–235, April 1962.

[39] Michael Kishinevsky, Alex Kondratyev, Alexander Taubin, and Victor Varshavsky. Analysis and identification of speed-independent circuits on an event model. *Formal Methods in System Design*, -(-):–, - 1993. Special Issue on Design Correct Circuits (To appear).

[40] Donald E. Knuth. *The Art of Computer Programming*. Addison Wesley, 1973.

[41] H.T. Kung and C.E. Leiserson. Systolic arrays (for VLSI). In I.S. Duff and G.W. Stewart, editors, *Sparse Matrix Proceedings*, pages 256–282. SIAM, 1978.

[42] S.Y. Kung. *VLSI Array Processors*. Prentice Hall, 1988.

[43] Leslie Lamport. Specifying concurrent program modules. *ACM Transactions on Programming Languages and Systems*, 5(2):190–222, April 1983.

[44] Leslie Lamport and Martin Abadi. The existence of refinement mappings. *Theoretical Computer Science*, 2(82):253–284, 1991.

[45] Leslie Lamport and Fred B. Schneider. Pretending atomicity. Technical Report 44, Digital Systems Research Center, 130 Lytton Avenue, Palo Alto, CA 94301, 1989.

[46] Leonard R. Marino. General theory of metastable operation. *IEEE Transactions on Computers*, 30(2):107–115, February 1981.

[47] Alain J. Martin. Compiling communicating processes into delay-insensitive VLSI circuits. *Distributed Computing*, 1(4):226–234, 1986.

[48] Alain J. Martin et al. The design of an asynchronous microprocessor. In Charles L. Seitz, editor, *Decennial Caltech Conference on Research in VLSI*, pages 351–357. MIT Press, 1989.

[49] David May. Compiling Occam into silicon. In C.A.R. Hoare, editor, *Developments in Concurrency and Communication*, pages 87–106. Addison-Wesley, 1990.

[50] Carver Mead and Lynn Conway. *Introduction to VLSI Systems*. Addison Wesley, 1979.

[51] Niels Mellergaard. PhD thesis, Department of Computer Science, Technical University of Denmark, 1993.

[52] C.E. Molnar, T.P. Fang, and F.U. Rosenberger. Synthesis of delay-insensitive modules. In H. Fuchs, editor, *Proceedings of the 1985 Chapel Hill Conference on VLSI*. Computer Science Press, 1985.

[53] J.D. Morison et al. ELLA: Hardware description or specification. In *Proceedings of 1984 IEEE ICCAD*, 1984.

[54] David E. Muller and W. S. Bartky. A theory of asynchronous circuits. In *Proceedings of an International Symposium on the Theory of Switching*, pages 204–243. Harvard University Press, April 1959.

[55] Larry C. Paulson. The foundation of a generic theorem prover. Technical Report Technical Report No. 130, University of Cambridge Computer Laboratory, 1988.

[56] Prunier. *Madame Pruniers Fish Cookery Book*. Danish translation by Gyldendal, 1966.

[57] Jerzy Rozenblit and Klaus Buchenrieder, editors. *Codesign: Computer-Aided Software/Hardware Engineering*. IEEE Press, 1993.

[58] Jørgen Staunstrup. *Specification, Verification and Implementation of Concurrent Programs*. PhD thesis, University of Southern California, Los Angeles, May 1978.

[59] Jørgen Staunstrup and Mark Greenstreet. Synchronized Transitions. In Jørgen Staunstrup, editor, *Formal Methods for VLSI Design*, pages 71–128. North-Holland/Elsevier, 1990.

[60] Jørgen Staunstrup and Niels Mellergaard. Localized verification of modular designs. submitted for publication, 1993.

[61] Ivan E. Sutherland. Micropipelines. *Communications of the ACM*, 32(6):720–738, June 1989.

[62] J.T. Udding. *Classification and Composition of Delay-Insensitive Circuits*. PhD thesis, Eindhoven University of Technology, 1984.

[63] V. Varshavsky, M. Kishinevsky, V. Marakhovsky, V Peschansky, L. Rosenblum, A. Taubin, and B Tzirliyn, editors. *Self-Timed Control of Concurrent Processes: The Design of Aperiodic Logical Circuits in Computers and Discrete Systems*. Kluwer Academic Publishers, Dordrecht, The Netherlands, 1990.

[64] Hans Henrik Løvengreen and Jørgen Staunstrup. Synchronous realization of asynchronous computations. In V. Stavridou and T. Melham, editors, *Proceedings of the IFIP WG 10.2 International Conference on Theorem Provers in Circuit Design: Theory, Practice and Experience*, pages 95–110. Elsevier, 1992.

[65] T. Verhoeff. Delay-insensitive codes – an overview. *Distributed Computing*, 3(1):1–8, 1988.

[66] S. Warshall. A theorem on boolean matrices. *Journal of the ACM*, 9:11–12, 1962.

INDEX